NFUScotland

The first 100 years
The story of NFU Scotland

By Andrew Arbuckle

Copyright © NFU Scotland

All rights reserved. No part of this book may be reproduced or transmitted in any form or by any means, electronic or mechanical, including photocopying, recording or by any information storage and retrieval system, without permission from the author in writing.

This book is sold subject to condition that it shall not in any way be lent, resold, hired out, or otherwise circulated without the prior written consent of the publisher, in any form of binding or cover other than that in which it is published.

First published in Great Britain in 2014 by

NFU Scotland
Rural Centre, West Mains
Ingliston
Midlothian
EH28 8LT

ISBN: 978-0-9928090-0-3

Design and layout: Karen Carruth

Printed by:
J Thomson Colour Printers
14 Carnoustie Place
Glasgow
G5 8PB

Contents

Message from the President I
Comment from Chief Executive II
Message from the Mutual .... III – V
A Cooperative view VI
The Next Generation VII
Preface IX

The Formative Years 1
In the beginning; First meeting; First internal moves; Early issues

The Hungry 1930s 11
Post War High; First Protests; Income Tax Arrives, Moves into Marketing, Labour Problem

The War Years 27
Surviving Depression; Summer Time, Sugar Beet; Labour Shortage; War's End

Price Review Years 40
Mechanisation; Harry Munro; Marketing Board; Fruit and Vegetable issues, Rabbits

The Arrival of the CAP 53
Import and Export Problems; Green Pound

Quotas, Set Aside, Food Promotion and the Environment 64
Food Quality and Promotion; Chernobyl and Currie; Garden Festival; Forestry; Islay Geese

Animal Diseases 76
TB; Foot and Mouth; Swine Fever; BSE; Aujesky's; Sheep Scab; Fowl Pest; Brucellosis

Weather 87
Snow; Wind, Drought; Flooding

Branches, Areas and Regions 93
Argyll and Islands; Ayr; Dumfries and Galloway; East Central; Forth and Clyde; Highland; Lothians and Borders; Orkney and Shetland; North East

Red Tape and Supermarket Power 114
Purchasing Power; Red Tape; Scottish Legislation; Demise of Marketing Boards; CAP changes

Union as an Organisation 124
Union Moves; Social Side; Publicity; Internal Organisation

Union Leaders 130

Acknowledgements 134

From the president
Nigel Miller

In 1913, the pressures on milk producers brought farmers together to form a Union. Scotland's farms were on the fringes of an economy driven by the Empire and focused on cheap imports.

Within months of the Union's formation, the vicious trauma of the Great War cut into the shipping lanes and transformed food into a precious commodity. Farming was suddenly at the centre of the war effort. Unfortunately, the imperative of food production melted away during the 1920s. Food prices plunged and land lay idle. A hundred years on, food security is again on the agenda worldwide. At home, a food and drink strategy is seen as one of the cornerstones of Scotland's economy.

The demands of new world economies have not only elevated commodity prices to a new level but created a wider appreciation of the quality and provenance that Scotland provides. Our industry has never had more exciting opportunities.

Beneath the positive market drivers, Scottish farming is on the edge of change. Europe has dominated the fortunes of agriculture since the 1914-18 war. Through more than 50 years of the Common Agricultural Policy, it has moulded the shape of farming, the countryside and our industry's regulatory framework.

Increasingly Europe is also defining the wider trading environment. EU Trade agreements with Canada and the USA will not leave agriculture untouched. Producers and their organisations must now proactively exploit these new markets to grow our sector and counter import threats.

As the Union enters its 101st year, the outcome of the Scottish Independence referendum in 2014 (and the potential for a future referendum on European membership) shape the political skyline.

We are also in the midst of a protracted CAP reform process. It is easy to pinpoint the absurdities of the EU but it is also too easy to forget that the EU has sheltered us from some domestic politicians that had no vision beyond the city of London. Throughout these austere times, the EU delivered €600m per year directly into Scottish farming.

Perhaps Scotland and the UK can learn from the Irish and place ourselves at the centre, not on the side-lines, if we are to drive positive changes in Europe. The Union, with its dedicated resource in Brussels, must be prepared to be a growing part of that process while also forging links beyond Europe as the reach of Scottish production grows.

At home, the reform of Agricultural Holdings legislation will change the landscape of Scottish farming and can open opportunities for a new generation. The Grocery Adjudicator creates new balance in the market, greater collaboration and innovation can grow the Scottish brand. Science, precision technology and smarter regulation must be part of future production.

NFU Scotland will face these and many more challenges over the next 100 years. The Union must be at the centre of all developments that touch the rural economy including renewable energy and carbon efficiency. The Union must also speak and work with the wider community while remaining at the side of every member with support, information and advice. It's never been more important to work together.

Comment from Chief Executive

Scott Walker

It is 100 years since a group of farmers from the west of Scotland came together with a vision of giving Scottish farmers a unified voice. From that meeting, NFU Scotland was born and for the past 100 years we have been a strong, clear and unified voice for the farmers, crofters and growers of Scotland.

I am proud, through the pages of this book, to be celebrating the past and I am looking forward to our second century and the role that NFU Scotland will have in shaping the future of farming in Scotland.

Three themes have been constant over the past 100 years and will continue to shape us in our next century. Politics, the market and assisting individual members.

Politics has had and continues to have a big influence on farming. Food security drove agricultural policy in the UK during the Union's early years and successive Governments encouraged farmers to produce more and, in negotiations with the Ministry of Agriculture, we had a role to play in fixing the support price to be paid for each agricultural product. Of course that situation no longer exists and while we no longer set prices alongside Government, the influence of politics is still strongly felt and the need to influence the decisions to be made by politicians is just as strong as ever.

But now, instead of operating on one front with a single UK Government, NFU Scotland must operate on multiple fronts in Holyrood, Westminster and Brussels.

From a pioneering role in the creation and establishment of a red meat promotion body to the development of farm assurance NFU Scotland has examined and driven new ways to ensure that Scottish produce is best positioned in the market. Our members are the bedrock of a hugely successful Scottish food and drinks industry and in shaping the future of farming in Scotland NFU Scotland will develop stronger links to those major end users – such as processors, retailers or drinks companies – who buy the produce that comes from our farms.

We are a membership organisation and alongside the work we do for the industry we have always provided help and assistance to individual members. This work can be the work that is most valued by the individual member and often the most rewarding for individual members of staff.

Few of the issues dealt with by NFU Scotland have a clear and simple solution and our members don't have a single point of view on many things. But the strength of the Union is that we can bring people together who have strong opposing views and we hold these debates and discussion. And after everyone has said their piece, we go forward with purpose and a direction to deliver a solution.

I thank all the members who have supported NFU Scotland during the last 100 years and all the people who have worked for NFU Scotland. When all is said and done this organisation is about people. The members who support us and the people who work for us.

Message from the Mutual
Richard Percy, Chairman of NFU Mutual Board

It could be seen as a play on words describing the links between the National Farmers Union of Scotland and the NFU Mutual Insurance Society as being of mutual benefit.

But for most of the past century, the two organisations have worked together with the 8,500 strong membership of the Union benefitting from the insurance policies of the Mutual.

NFU Mutual also provides financial support for the Union which enables both organisations to provide a local service network.

It also enables the NFUS to provide services to members without reflecting the full cost in subscriptions.

It was almost on the tenth anniversary of the start of the Union in December 1922, that the Insurance Sub-Committee minutes report talks had taken place with the National Farmers' Union Mutual Insurance Society Ltd.

The outcome of these discussions was that the Mutual, set up at the beginning of the century by seven Warwickshire farmers, would extend its insurance scheme to Scotland by becoming the official insurer for the Union.

Only twelve months later, the Union's annual report recorded "one of the outstanding features of the year has been the success which has attended the inauguration of the Insurance Scheme.

"A large number of members have taken advantage of the special terms offered, and the volume of business done in Scotland has exceeded all expectations. Claims are settled promptly, and there is ample security for all policyholders."

One of the earliest policies was taken out by James Duffus, Millfield, Turriff, who insured all of his belongings from growing crops and livestock kept on the farm through to household goods. As can possibly be deciphered, the premium for this coverage was less than £5.

Initially a room borrowed from the NFU became the first Glasgow office of NFU Mutual but the rapid growth of the business soon made it necessary to move further along Vincent Street.

Larger premises were acquired over the years until 1975 when the Mutual moved into a purpose-built office in Bath Street. Then in 2011, NFU Mutual's Glasgow operations moved to Centenary House in the heart of the City's financial district.

Back in the early days, at the 1926 annual meeting, it was recorded that NFU Mutual business among members in Scotland had increased by 40% and two years later for those still unconvinced about the link between the Union and the Mutual,

Ross Taylor, president, said while discussing a rise in Union subscriptions: "Mr Purser (NFU Mutual manager in Scotland) will tell you that even if they put their motor cars through the NFU Mutual, they would in most cases get the amount of their subscription back."

Since those early days, the policies of NFU Mutual have moved with the times as agriculture developed. As tractors replaced horses and technological advances enabled productivity to rise dramatically, new insurance policies were brought in.

The Second World War brought a whole range of new problems for farming but the NFU Mutual reacted with policies associated with war time including the loss of livestock on account of bomb damage.

The post war years saw the extension into Scotland of the full time agent network which had been pioneered down south and by 1962, Union Organisation's committee convener, WJC Little, praised the Mutual for growing into a business with assets in excess of £27m and a gross income from premiums and investments of over £10m per annum. "That must give encouragement to the groups of farmers embarking today on marketing and other co-operative ventures," he remarked.

T+he 1960s also saw the NFU Mutual having to deal with its first major animal disease outbreak when a foot and mouth virus saw more than 100 farms in Scotland and many more in England having to slaughter their livestock. This sad experience was replicated in the 2001 epidemic and to a lesser extent in 2007.

Just how Mr Little would describe the £13 billion the Mutual now have in assets is difficult to imagine. Equally the £1 billion it now collects annually in premia is a far, far cry from the first year figures which showed a total income of £311 and a profit of £16.

For 102 years, NFU Mutual has provided the insurance services farmers need while standing shoulder to shoulder with the NFU to ensure that they are represented by a strong organisation.

As 'The Mutual', with farmers' interests at its heart, NFU Mutual does far more than provide protection and safety through insurance, pensions and investments. It also provides 'Mutual Support' to farmers and their families in a number of ways:

■ We help farmers fight rural crime, and tractor theft in particular, through financial support for the police vehicle crime unit and sponsorship of rural crime conferences. We also run the NFU Mutual's Country Crimefighters Awards which celebrate the best rural security initiatives of local groups, individuals and the police.

■ We are using our expertise built up over a century insuring farmers to help them work safely and reduce the toll of accidents on Scotland's farms. In addition, NFU Mutual's Charitable Trust supports a wide range of agricultural causes:

In 2012 The Trust made donations totalling £266,000 to a range of organisations which provide invaluable support to farming families across the UK.

In addition to these donations, the Charitable Trust made a major contribution of £150,000 to the Prince's Countryside Fund to help farmers who are struggling as a result of the appalling wet weather in 2012.

Looking to the future of British Farming, the Charitable Trust's Centenary Award has provided support for 20 post-graduate agricultural students from across the UK.

"Over its first 100 years supporting Scotland's farmers, the National Farmers Union of Scotland has done a magnificent job campaigning for the nation's farmers."

"NFU Mutual is proud of our long and successful partnership which continues to work extremely well for both organisations and, most importantly, our members.

"Farming remains at the heart of our business. From our formation in 1910 by seven Warwickshire farmers, NFU Mutual has grown to become a business widely acknowledged for its integrity, its financial strength and its customer service.

"As a farmer myself, albeit from south of the border, I firmly believe that a strong relationship between farmers, NFU Mutual and the NFU is in our mutual interest and hugely important to the future fortunes of farming families."

Throughout the NFU Mutual's history, Scotland has been well represented at Board level. Currently former Union president, Jim McLaren sits on the Board and back in the 1980s, Morgan Milne followed his stint as Union president by chairing the Mutual's Board.

Foot note or football note

The NFU Mutual's support has not all been in insurance, as in 1996 the head office at Stratford-upon-Avon was used by the Scottish Football team as a base for their preparations for Euro96. This was the last time the Scottish football team qualified for an international competition.

The choice was made because the Mutual's sports club had a top quality football pitch, together with extensive recreational facilities including an indoor swimming pool.

A Cooperative View

James Graham, SAOS Chief Executive

It's an honour to be invited to contribute to NFU Scotland's Centenary book. The Union is held in such enormous respect by everyone with any connection to farming, and that of course is down to a century of effective representation delivering real results on behalf of its members and the wider industry. Some of today's buzz words certainly apply.

The Union is the authentic 'voice' of Scottish farming, responsive and resilient, in fact a 'rock' for farmers who have over the years had to engage with increasing political interest and intervention that has sometimes been requested and required, and at other times not been welcomed.

A notable strength of the Union has long been its ability to mobilise exceptional leadership in a crisis. Time and again, the Union has enabled gifted leaders to emerge who have made outstanding contributions on behalf of fellow farmers in extraordinarily difficult situations.

It's admirable that any organisation should be able to nurture and develop leaders. NFU Scotland has proven over a century that it can do so, and repeatedly. The Union's stature and influence look secure and set to further extend in a shrinking world so long as it continues to be a platform and enabler of new leaders.

I am bound to say that the Union serves as an excellent example of the advantages and gains available through collective farmer action. Both SAOS and NFU Scotland serve the interests of farmers who recognise and acknowledge the potential of working together – NFU Scotland in the political arena, SAOS in the commercial arena.

The organisations complement each other, have an open and continuous dialogue, and a good working relationship that has endured for a century. In SAOS, our 70 agricultural co-op members generated throughput of £3billion in 2012, proving that working together is as potent for farmers commercially as it is politically, enabling them to access the commercial benefits of scale whilst retaining their independent family farms.

We thank NFU Scotland for its partnership and support over the last 100 years, and look forward to achieving more together as Scotland's farmers face ever greater challenges in a world of opportunities in the next 100 years.

The Next Generation
Penny Montgomerie, SAYFC Chief Executive

The NFU Scotland has played an integral part in the current and future personal development of our members, and enables the association to remain in touch with the agricultural sector. With more than 75% of Scotland's land mass under current agricultural production, the union is as important as ever in ensuring our members have a voice and a range of opportunities that allow them to progress with their goals and aspirations.

Our unique organisation led by the members for the members, has encouraged individuals to think and speak for themselves, and as the association evolves the original young farmers motto created in the 1950's still remains relevant today "Better Farmers, Better Countrymen, Better Citizens". NFUS enhances and supports this ethos, inviting members to attend meetings, approaching them for their views and working with the SAYFC National Office to build a network that welcomes young farmers from throughout Scotland.

Throughout our 75 years, the union has supported a number of events which has allowed both parties the opportunity to raise their awareness with our members and the general public alike. Without their support some of these competitions would not have taken place.

Our members come from a range of backgrounds which NFUS understands and appreciates the need to deliver a range of different data depending on their interests. They ensure information is well balanced with both policy and case studies, taking our members opinions to those who have the ability to make a change. More recently members have been invited to attend the NFUS New Generation Roadshow which allowed them to gain a better understanding of current changes to the agricultural industry in a comfortable environment where they are with like minded individuals. This shows how adaptable the union is and its superb understanding of the needs and requirements of those who one day will be rural leaders, and indeed potential NFUS members or staff.

The Scottish Association of Young Farmers is very proud to have formed a strong relationship with NFUS, and is delighted to be invited to congratulate them on their centenary. The association looks forward to working to develop the relationship even further over the next 100 years, and sends its best wishes for the future success. Congratulations NFUS!

Preface

I know it is a cliché but it is impossible to compress the one hundred years of history of the National Farmers Union of Scotland into one relatively slim volume. The range of work this organisation has covered over the past ten decades means that not all of them can be covered.

Therefore, my aim has been to highlight the main issues and challenges faced by the Union. The book follows a rough chronological path but rather than have issues overlap into other chapters, they have generally been dealt with in the era when they were most relevant.

There are several style points to note. Throughout the book, NFU Scotland which was originally called the Farmers Union of Scotland and which was briefly called 'the Scottish Farming Union and the Chamber of Agriculture for Scotland' is referred to as the Union. This is not my christening merely an attempt at simplicity.

Andrew Arbuckle

The past century has seen this country adopt changes in both currency and weights and measures. While the older generation might still be able to convert the old standards into today's ones, the book has, with one exception, provided the conversion. The exception to this rule has been changing old tons to modern tonnes with the difference being so slight as to be immaterial for the reader.

This book would not have been completed without the invaluable help of my brother, John Arbuckle. Together, we pored through dusty minutes and museum archives in order to pick out the history of the Union and together we travelled over the country to meet people with links to the organisation.

There has been tremendous support from a great number of people within the Union. Members, Secretaries, office bearers - both past and present - have been universally helpful.

If there are errors within the volume, these are mine alone. If there are omissions, I apologise.

With these caveats, I sincerely hope you enjoy reading a little of the history of the foremost rural lobbying organisation in Scotland.

Andrew Arbuckle

The Formative Years

The background
Apart from the competitions in the livestock rings, agricultural shows have traditionally served another purpose for those involved in farming: they provide a meeting place where the issues, problems, concerns and even just downright gossip are exchanged.

Thus it was at the Highland Show, held in the first week of July 1913 in the 45-acre St James Park, Paisley. The crowds critically watched the stock judging, especially the draught horse classes where there were some 84 Clydesdale stallions forward, strongly indicating the show was being held in the heartland of the breed and that the original horse power was still the main engine in agriculture.

They also watched the record entry of 124 Ayrshire cattle, very much the local breed vastly outnumbering the seven Jersey cows, twelve Shetland dairy cows and two dozen British Holsteins; the last being the newcomers to the show ring.

The sheep pens were dominated by Blackfaces and Border Leicesters with smaller numbers of Shropshire, Suffolks and Oxford Downs.

But with the judging completed the talk was of the increased imports of produce from Europe and the Dominions; in fact from all over the world as productive farmland in new countries was cropped and stocked for the first time.

Although the visitors to the Paisley show would not have known the import statistics at that time, they would have felt the effects in the prices offered for their grain and fat cattle, sheep and pigs as well as dairy produce. Britain at that time was importing one third of its beef and mutton requirements from North and South America and Australasia. And it was not all arriving on the hook, as more than 150,000 head of live cattle were brought over from America that year.

Further importations came with three quarters of the country's pork and bacon from Denmark and the Netherlands, butter and eggs from Russia and even rabbits from Australia, New Zealand and Belgium.

Grain flowed in from around the globe, undercutting home markets with Canada, the USA and Russia heading the list of countries trading their farm produce for trains, ships and other manufactured goods from Britain. Four fifths of the country's wheat was imported in the first decade of last century.

Wool was also being imported in large quantities from both New Zealand and Australia to feed the textile mills and these imports were undercutting the one shilling (5 pence) per pound Scottish farmers were getting for their wool.

Some of the older show goers could say that the price of sheep was less than it had been fifty years previously; a comment that would have equally applied to other commodities, such as beef and grain.

In fact, from 1870 farming had been on a downturn as Britain embraced a free trade policy which not only opened the doors to imports but provided preferential railway rates for foreign produce once it had arrived in the ports.

By 1914 it was reckoned that two thirds of the country's food was coming in from abroad. In contrast, Germany which had operated a protectionist policy for its farmers, was almost self-sufficient in food -an important factor given what was ahead.

Farming was in a bad way

In the first decade of the last century most of the farmland in Scotland was tenanted but the lack of profitability prompted many tenants simply to abscond at some point through their tack (lease) of seven years, leaving the landowner to find another farmer to take over what was inevitably a rundown unit. Throughout the country, land which had been in crop or under improved grassland reverted to scrub and gorse

There were other problems facing the industry. Many small dairy producers were selling milk to operators in the towns and cities and were dissatisfied with the reward they were getting for their work and effort.

Dairy farmers claimed the buyers bought quality and sold quantity and the primary producers were the losers in this situation.

"What is to be done?" the greybeards in the industry must have thought as they met at the show. There had been earlier similar discussions at Kilmarnock auction market on the same themes.

Further evidence of the distressed state of agriculture in those years emerged in the raft of letters from angry and disillusioned farmers in the pages of both the Scottish Farmer and the North British Agriculturist - the two trade magazines of the day.

Some of the anger was directed at the Scottish Chamber of Agriculture, a long established organisation with a focus on improving husbandry and developing sporting enterprises. This was the era of Victorian obsession with deer and grouse moors and the SCA with a membership mainly of landowners and larger scale tenants had an interest in these aspects of rural life as it helped preserve its wealthy supporters.

But such an agenda cut little ice with working farmers looking for help with day-to-day issues such as dealing with local authority levies for road repairs, education levies, wayleave payments and a range of problems arising from the railway companies with a virtual monopoly on transport.

There was no need for an organisation to help education in new farming techniques as three colleges of agriculture, the West, East and North, had all been established in Scotland in the first decade of the twentieth century.

Equally there was no need for an organisation to help with the marketing of produce as the Scottish Agricultural Organisation Society had been set up in 1905 following the establishment of a number of local co-operatives, including Tarff Valley, at Castle Douglas which operates successfully to this day.

But there was a need for the farming industry to be represented both at local and national level to deal with issues such as rabbit trapping, which was not permitted on some let farms and major issues such as the controversial introduction of National Insurance.

There is no doubt that someone either at the Highland show or at one of the earlier market meetings must have mentioned the establishment of the Farmers' Union in England four years previously and how this organisation, with a massive membership of 35,000, was already making its presence felt in the industry.

Under the presidency of Colin Campbell, who was one of thousands of Scottish farmers who had moved down to the sweeter, more easily worked land in Lincolnshire in the early years of the last century, this young organisation had created a list of priorities that included branding of all imported produce as foreign and the strict adherence to regulations concerning the slaughter of all 'foreign and colonial sheep and swine' at ports of entry to prevent disease entering the country.

And so the worthies at Paisley show decided to hold a meeting in the coming months to see if a similar organisation could be set up in Scotland.

From the minute books…

29/11/1917 Dalry Branch

With the officer of the Agricultural Executive demanding the ploughing up of one third of all farms, concern was expressed there would be a great shortage of grass for young dairy stock. Members felt the officer was not making sufficient allowance of land unsuitable for ploughing and that other parts of the country were being dealt with more leniently.

29/11/1917 Dundonald Branch

Members meeting at the Old Castle Hotel recommended that ploughman's wages should be either £2 without perquisites or 35 shillings(£1.75) with perks.

The first meeting

They came from all parts of the West of Scotland and more came down from the North East of the country, many of the former having milked their cows in the morning and many of the latter having foregone a day trying to bring in their cereal and potato harvests.

They came by train from where they made the short journeys from Queen St and Central stations along to the Religious Institution Rooms, 200 Buchanan Street, Glasgow. The Rooms are still there but today they are known as the Buchanan Galleries.

There they were met by other farmers who arrived in motor cars which were then just making their presence felt in city centres. Many of those who came to this fateful meeting on the first of October 1913 were representing so-called improving agricultural societies that recognised the fundamental weakness of a farming industry not having a national voice.

Among those who made a longer trip than most was Alfred Churches, a tenant farmer from Kent and more significantly, a vice president of the Farmers Union of England.

Such was the crowd at the meeting, The Scottish Farmer reported that people were standing not only in the hall, but on the stairway leading up to it.

Churches had been asked to speak on the benefits of farmers working together and in promoting this theme he referred to the English Union's motto "defence not defiance." He also struck a chord with his audience when he claimed, "We try to help every farmer in trouble but our main difficulty is that although farmers produce the nation's food, we do not receive a fair deal in legislation."

He added that members of Parliament did not listen to farmers because all they heard were individual voices which were not unified in their concerns.

"Unless you support each other, then agriculture will go under!" he thundered. Together, he said they could tackle the railway companies who at that time were

---

A convention of

West of Scotland Dairy Farmers

and others
interested in agricultural questions

will be held in

The Religious Institution Rooms

200 Buchanan Street, Glasgow

on

Wednesday, 1st October, 1913

at 2 p.m.

All farmers cordially invited.

responsible for almost every item which came onto or left the farms; this mode of transport being especially critical for milk producers who needed to get their milk to market every day.

Churches also claimed that some local authorities needed farmers to work with them to help eradicate tuberculosis which was killing many young children at that time and he asked the crowd, "Is that alone not a reason to work together?"

The chairman of the meeting, sensing the mood of the audience, took up the clarion call. "We are the butt of Government, of public bodies and of commerce. We are legitimate game," going on to say that position needed to change and today was the day to start the process.

Churches may have had an ulterior motive in his visit to Scotland - to get any new Scottish farmers' union to work with his organisation.

A formal approach to that effect was made some six months later but was rejected. More of that anon.

> **THE FIRST FOUR RESOLUTIONS**
>
> 1. That this meeting of farmers and others resolves and pledges itself to form a Union, with the object of mutual help and the furthering of the interests of Farmers in general and Dairy Farmers in particular, to be called "THE FARMERS' UNION."
>
> 2. That a Committee be formed of Delegates from the various districts to draw up Rules and Regulations for the guidance and control of the Farmers' Union; to appoint a President and Secretary and other officials as may be deemed necessary to carry on this movement for united action
>
> 3. That each Member and Delegate undertake in his respective district to foster, encourage, and extend at every opportunity and by all lawful means the idea or principles of Union, and of the immense benefits which must follow from united effort.
>
> 4. That a small entrance fee of 5/- be subscribed by each Member on joining to provide funds for carrying on the work.

Whether it was Churches' evangelism or the pent-up mood of those present, the decision to set up a farmers' union was supported unanimously.

Four resolutions were then put to the meeting. The first and most important which reflects the original driving force behind farmers uniting was "that this meeting of farmers and others resolves and pledges itself to form a Union with the object of mutual help and the furthering of the interests of farmers in general and dairy farmers in particular to be called The Farmers' Union."

That name, with the addition of Scottish to differentiate it from the English Union was retained until 1947 when the then president Ian M Campbell proposed the title, the National Farmers Union of Scotland.

Churches also advised the meeting to be independent of any political party; a Scottish Union policy which holds to this day, although particularly in its early days, some branches such as Turriff supported local candidates. But unlike the English Union with Sir Reginald Dorman Smith (1936/37), the Scottish Union has not provided a Minister of Agriculture.

At this point in the inaugural meeting, those attending elected as their president, William Donald, Fardalehill, Kilmarnock, a farmer but also significantly the owner of the Kilmarnock auction market where the first informal meetings had been held. From his election that day Donald guided the new organisation through its formative years and by the time he stood down in 1919, it was a national organisation with 6,000 members.

Incidentally Campbell, the Scots born leader of the English Union held office from 1909 to 1917; a length of service only bettered by Jim Turner, later Lord Netherthorpe, who served from 1945 to 1960 and Henry Plumb who served from 1970 to 1979.

Such was the mood of optimism and future stability, the meeting also appointed a permanent secretary, a Mr A Sturgeon from East Grinstead in Sussex. But with due Scottish caution, his starting salary was £3 per week for the trial period of two months during which he could also claim third class train fares.

William Donald, of Fardalehill, Kilmarnock – the first president of the NFUS

In the midst of the euphoria, there were doubters, who in the following weeks in the agricultural press, expressed warnings that farmers had never previously united in their concerns and the Union would surely falter.

Reporting on the meeting, The Scottish Farmer admitted the Farmers Union had made a great start but then qualified its comment. "The zeal of those who overcrowded the meeting was all right provided it was directed into a profitable channel."

The paper also noted that all previous efforts at unison among farmers had "shipwrecked on the snag of calling for co-operation and united action and then allowing the other man to co-operate and take the action." The trumpet call to those present was to "stand loyal and true to one another."

The next internal moves

By the time of the first annual meeting which was held in the Masonic Halls in Regent St, Glasgow, on the 25th of February 1914, the fiery cross had crossed the country and 109 members attended. They came not just from the West and South West but, as J Montgomerie Pearson, of Over Lethame, Strathaven the minute taker noted, also from Stirling and West Fifeshire. The delegates heard President Donald predict that within a year the Union would be a national organisation.

It was at this meeting the Union agreed its operational framework of branches that fed into Area or County committees and then into the National Council. The early annual general meetings were thereafter dominated by reports from the various areas and the issues being dealt with at that level.

At the same meeting the Union appointed its first auditor, Joseph Allison, who went on to audit the books of the organisation for the next fifty years.

By 1918, the Union, faced with problems across the farming spectrum, had set up committees to deal with the wide range of challenges faced by the ever growing membership.

The committees were: Insurance, Beef and Mutton, Milk which is the only committee to survive unaltered for one hundred years, Parliamentary Representation, Constitution and Rules, Local Taxation, Diseases of Animals, Milk Standards, Minimum Wages, District committees, Game Laws and Heather Burning, Finance and Trading

Joseph Allison, auditor for the NFU for the first fifty years of its existence

Although the original annual subscription had been agreed at five shillings (25 pence), delegates unanimously supported a resolution that this should be doubled to ten shillings (50 pence). This was the first and only time so far that the union has doubled the annual sub, although twice in the 1980s, a 20% increase was adopted in order to keep pace with raging inflation.

However, the delegates were not so amenable to the proposal in 1915 to affiliate to the English Union. The breaking point was the one shilling (five pence) per member affiliation fee that the English Union requested.

The potential link up did not die there as a further move was made in 1916. The Union minute book relates: "The views expressed by the English delegates (on the subject of a potential link up) were reported. Particularly the proposal by the English secretary to the effect the Scottish Union should join England as a branch organisation.

"While this proposal could not be entertained, it has been agreed that the two secretaries should act in close cooperation in all matters of interest and this is now being done."

The first issues dealt with by the Union

With its roots deep in the milk producing areas of Scotland, it is hardly surprising that the very first motion to the newly formed Union related something having to be done about the poor price being offered for milk. It was submitted by a Mr Frood and was passed unanimously.

This was followed by other resolutions from dairy farmers in the West and South West of Scotland to the Central Executive which was the main decision-making body of the union.

These related to various problems from the stopping of the sale of 'warm milk' which, because it went 'off 'quickly, jeopardising the healthy reputation of milk to the 'iniquitous' rail charges for milk. In those days the railways offered the only feasible transport to the big cities and there were always suspicions the rail companies were profiteering with their charge of a penny (0.4p) a gallon haulage all milk.

Outwith the milk producing areas, other branches raised their concerns over the fact they were being held to ransom by the rail companies intent, as the farmers saw it, on making profits rather than feeding the nation.

Issues ranged from demurrage charges levied when farmers ordered trucks and did not use them right away, through to problems with the hiring of railway sacks. These were owned by the railway companies and were used for transporting grain. They were made of thickly woven jute but as farmers know, rats and mice are no respecters of such material if grain is available; the net result being many claims of sacks with holes being delivered balanced by many claims of sacks with holes being returned.

Although there was no reference to the looming clouds of war in the early days of the Union, its life and workload were shaped by the First World War (1914-1918) which broke out less than a year after the inaugural meeting.

Before the war was over, four years later, one quarter of the farm workforce had gone off to fight for king and country. Unlike those working in the mines, shipyards and munitions factories, farming was not designated as a reserved occupation.

Such was the shortage of labour on the farms that one of the first national issues raised by the Union was a bid to reduce the school leaving age. They were

From the minute books...

29/11/1917 Beith Branch
Matthew Gilmour, secretary reported the meeting was addressed by A W Sturgeon the national organising secretary on the need among farmers to work together in order to get better prices for milk and other farm produce as well as obtaining fair treatment in Parliamentary legislation.

13/7/1917 East Lothian Branch
Members called on the Government to rescind the sale of Horses Act which they described as mischievous, unfair and oppressive in that it allowed the Army Remount Department to commandeer work horses for much less than their real value.

unsuccessful in this as they were defeated by the educational lobby.

The Union also had to deal with a requirement that farm staffs had to do four hours of "drill work" in case of an enemy invasion. This was the First World War version of the Home Guard and it was a hefty enough imposition during the quiet periods on the farm but it was well nigh impossible during sowing time or harvest. A concession for these periods was obtained in some areas thanks to Union intervention.

In the first World War and, to a lesser extent, the second, one of the first demands on agriculture was the commandeering of horses

The lack of manpower was not helped by a simultaneous loss of horsepower as almost from the first shots in anger, tens of thousands of draught horses were commandeered from farms, no longer to pull ploughs and seeders but to a future of pulling guns, ammunition and other supplies to the front lines. The plundering of horses from Scottish farms was exacerbated when the military recognised the ability of Clydesdales to work in deep mud.

From being an importing nation, the country had to move as far towards self-sufficiency as possible as enemy gunships patrolled the oceans. Land which had reverted to grassland from the last boom time in 1870 was ploughed up and cropped. Aided by a ploughing-up subsidy and vastly increased prices for every commodity, there was almost a quarter more cropped land in 1918 than four years previously; this despite a reluctance by some farmers to reduce their grass acreage which was easier to maintain than cropped land with a much reduced workforce.

To help ensure maximum food production, County Agricultural Executives were established to ensure proper cultivation and the running of farms. They also had the authority to take over badly run units if necessary - a policy repeated in the Second World War.

Despite these and other efforts, by 1917 the country was almost starved into submission, a fact acknowledged by Prime Minister, Lloyd George who described famine as "the enemy within." Only drastic action such as 'meat free' days helped eke out food supplies in the latter years of the war.

Despite the controls placed on marketing, food shortages brought increased profitability to farming. Milk which had traded at 6 pence (2.4p) per gallon in 1913 was worth three shillings (15p) some four years later. Sheep, cattle and grain prices all doubled during the war years yet land rents remained the same because the vast majority of farms were on seven or fourteen year leases.

The Scottish Farmers' Union soon locked horns with longer established Scottish Farm Servants' Union which had its stronghold in the North East of the country. The redoubtable Joseph Duncan whose family were farm workers was the driving force behind the farm workers' claims for better wages and working conditions.

With no national wage agreements in those days, most deals were left to the areas of the Union to sort out. Shortages of workers and profitable farming in the latter years of the war saw workers' wages rise but housing and working hours were still harsh compared with today.

In 1919, the first national conference between the Union and the farm workers took place in Perth and at that meeting, decisions were made to limit the working hours of ploughmen. From then on, Saturday afternoons were no longer part of the working week.

In the afterglow of winning the war, there was optimism in agriculture. Union honorary president James Gardiner, MP, spoke of the need for everyone, farmers and farm workers alike to work together and use the comradeship learned in the war years for the benefit of the industry as a whole.

"If anything has been learned from the terrible war from which we have just emerged it was that selfishness must be eliminated. If we are going to consider our men, we must remember they are not housed in Scotland as they should be. If the Union is going to do anything they must make it more comfortable for men to work on farms and to induce them to take an interest in the work they do."

Milk deliveries in the early years of last century

The Hungry Years

The post war high and the rapid fall in profitability

As the cease-fire sounded to end the First World War, Scottish farmers were receiving far higher prices for their milk, beef, mutton and grain than they had ever done.

A shortage of labour and an increase in farmworkers' wages may have latterly eroded some of the extra cash being paid but overall, with a bumper harvest in 1918, the industry was extremely profitable.

Throughout the war years, the Government appointed County Agricultural Executives had strictly controlled agriculture. These bodies told farmers when they could market their cattle and sheep and sell their grain and potatoes. In return, farmers were guaranteed minimum prices that generally exceeded production costs by a considerable margin as the Government strove to ensure maximum home food production.

The Corn Production Act of 1917 had guaranteed grain prices up to 1922. However, in 1920 this was superceded by another Act that set the prices for wheat at 68 shillings per quarter (£15.32 per tonne) and oats at 46 shillings also per quarter (£15.10 per tonne). When the Act was passed, it was stated these prices were to be re-valued annually and farmers thought their future prosperity was assured.

But within two years of the ending of hostilities, food was not just being imported into Britain but some of it was also being subsidised to do so. As a result prices fell dramatically.

With the collapse of grain prices on the world market and with hundreds of thousands of hungry unemployed people to feed, the Government rushed through an Act repealing the previous one that guaranteed cereal prices.

Referring to this change in policy in the early 1920s, one Union commentator remarked: "once again we see the sacrifice of British agriculture to cheap food and free trade."

Trenchantly, the 1921 annual report of the Union comments: "The anticipated slump in prices has come about. After repeated declarations by the Government that they would not allow agriculture to slip back into its old state of neglect and stagnation, they have repudiated these measures. Scottish farmers must rely on their own capacity and resource. State aid and state interference can be dispensed with."

The adjustment for farmers from having every morsel of food they produced being bought at prices never experienced before to a point where potatoes were left to rot, where milk was poured down drains and wool was hardly worth taking off the sheep's back was difficult to take.

This was where the increasing national influence of the Union began to play a part. From dealing with many minor issues at a local level, as it had done in its early years, the next period from 1920 to 1939 saw the Union move to dealing with Parliament and shaping national legislation. That shift did not come easily and definitely did not come without considerable effort.

The bitterness of having responded manfully to the call for extra production during the 1914-1918 war years and then seeing the food they produced subsequently being used as a method of keeping an increasingly restless national workforce quiet features regularly in the minutes of the Union in the first half of the 1920s.

By mid decade, the economic position may not have improved greatly but at least by 1925, the Union president Alex Batchelor, a large scale dairy farmer and potato merchant based on the outskirts of Dundee claimed: "The period of depression since the boom years of the Great War would now appear to be in the process of passing away, and provided that there is no unnecessary interference by legislative action, and no unfair preferences given to our competitors overseas, whether Colonial or Foreign, one might almost be tempted to predict that the year now begun will witness a considerably improved position for all engaged in agriculture."

His optimism was not rewarded. Prices did stabilise but only at a low level until 1929 when they took another slump downward followed by yet another two disastrous years for the farming industry.

In 1929, Blackface lambs could be bought for 7 shillings (35 pence) per head. Potatoes were hardly worth lifting. Those who did often had to settle for £1 per ton. Wheat in the Edinburgh market was 20 shillings and sixpence a quarter (£8.92 per tonne) and

Alex Batchelor tried to inject some optimism in the 1920s

John Spiers complained about the importation of potatoes

oats 15 shillings and 9 pence (£5.26 per tonne). In country areas, grain prices were even lower.

Farming was in a poor state; good stock farms could be bought for £10 per acre and many tenanted places were left unlet.

That year saw the Central Executive of the Union lobby Government on two measures where it claimed reform was essential. "Firstly, the whole question of the importation of agricultural produce with special regard to malting barley, condensed milk, oatmeal and potatoes. Our country was self sufficient in these products and some measure of protection on imports was called for.

"Secondly, it should be made compulsory to mark distinctly all imported foodstuffs so that consumers may be able to distinguish between them and the home produced article."

A reader's letter to the Dundee Courier the following year pursued a similar theme on labelling imports. "One should not be able to sell beef as Scotch when the meat has been taken from an ox that has had a few Scotch turnips, some Scotch straw and some Bombay cake stuffed into it for a few weeks or months."

There was apparently a requirement that butchers had to notify their customers if meat was imported but Douglas Oliver from Jedburgh Branch was not convinced the requirement was being met.

"The Order clearly states that a notice should be placed in a prominent position in any shop where imported meat was sold. The police are not bothering themselves about the Order. In fact it might as well not exist." He wanted butchers either to sell home produced meat or foreign and then customers would know where to shop.

Under pressure from the Unions, both north and south of the Border, the Government made some concessions to the farming industry. Land and farm buildings were excluded from rates. Grants were also made available for drainage work on farms with the Government determined not to let land slip back into disuse.

Despite that aim, the acreage under grain fell by almost half compared to where it had been at the end of the First World War. As 'dog and stick' farming once again took over, the area under grassland rose by almost half a million acres.

As the country entered the 1930s, any memories of good times in farming were swept away with more and more imported food coming into the country.

The first protests organised by the Union

Surprisingly, given the origin of the Union in the milk industry, the first protests were over the importation of potatoes with Perth County in 1924 organising mass meetings of farmers to protest against these imports.

The issue was not a one season wonder as in 1927, president John Spier referred to the attitude of Government on the importation of potatoes, stating: "Potatoes are allowed to come in here even when the country of origin is known to be affected with various potato diseases. Compare that with the position taken by the USA government, which refuses to allow the importation of even one ton of potatoes from this country because certain areas are affected with wart disease."

The milk industry was not immune to imports and in 1930 the NFU issued the statement: "For our part we wish that Mr Baldwin (the UK Prime Minister) were a Mussolini (the Italian fascist leader who a decade later would lead his country into the Second World War) for he tells us if he were, he would be able to put an embargo on the importation of milk products into this country."

However, the first national show of strength by the Union came in the early 1930s with protest meetings held in all the main market towns across the country. The bone of contention was the importation of grain undercutting the home trade. Specifically, imported oats with a bounty or subsidy were coming in from Germany at 13 shillings (65 pence) per quarter (approx. £4 per tonne) with German producers, as noted by Dumfries area, receiving 10 shillings (50 pence) subsidy per quarter. Some 646,000 quarters of oats came in, a tonnage equal to 117,000 acres of crop. Not only did this undercut the depressed home trade but the country of origin of the oats was hard to take for those who had either fought in the Great War or who had lost relatives so recently fighting the Germans.

On the overall outlook for farming at that time, Walter Barrie, secretary of Selkirk County recorded in his report to National Council: "Prices for fat and store stock continue to tumble. Potatoes and oats are having to be given away at record low levels. There must be no ceasing to urge the measures necessary to give agriculture and the nation a chance. There never was more need to take to heart the warning of the poet, Oliver Goldsmith which was - A bold peasantry, their country's pride when once destroyed, can never be supplied."

Typical of the scale of the protests was the one in Angus, which involved a march from the Scott and Chalmers livestock market to the centre of Forfar. It was, according to the local paper, preceded by a pipe band with six or seven thousand people, including "farm servants who had been given the day off coming by special train, bus, push bike and even Shanks' pony to make their feelings known."

They were addressed by former Union president, Alex Batchelor who told the assembled throng: "the farming industry is fighting with its back to the wall. All we want is a fair deal."

From the minute books...

January, 1931 Dalry Branch

Patrick Comrie, secretary was instructed to write to headquarters suggesting that the Union should ask members of all Branches to discontinue the practice of making payments of Arles to servants, thus sealing a deal on future employment. This issue had arisen when, despite receiving this engagement money, some farm servants had failed to turn up for work.

A few weeks later, another large crowd of up to 6,000 people from "mansion house, farm and cott" according to the local paper, turned out at Perth market also complaining about the corn coming in from Germany.

In Aberdeen, the Union president Maitland Mackie addressed a full house at Pittodrie Park, Aberdeen football club's ground. He was joined on the platform by Joseph (Joe) Duncan, the Aberdeen-born driving force behind the Scottish Farm Servants' Union (SFSU).

Mackie was reported to have regarded that as some sort of ecumenical triumph, and perhaps it was, but Joe Duncan thought he had the best of it. He had been campaigning for a half day off on Saturdays and he had got the farmers to agree to let their staff off in order to demonstrate that Saturday afternoon. Thus proving work was not as essential as the farmers had been saying.

Further North, the Orkney Herald ran a large advertisement asking for farmers from all the islands to come to a mass meeting in Kirkwall to complain over the importation of oats from Germany.

Helpfully the paper listed the times for the various boats that would bring the protestors to the main town. On the day itself, some 5,000 farmers and farm workers met in the shadow of St Magnus Cathedral.

There they heard William Corrigall, Northbigging, Harray, call for a ban on the importation of the foreign 'bounty led' oats.

In March 1931, the Rhinns of Galloway Area organised their protest meeting

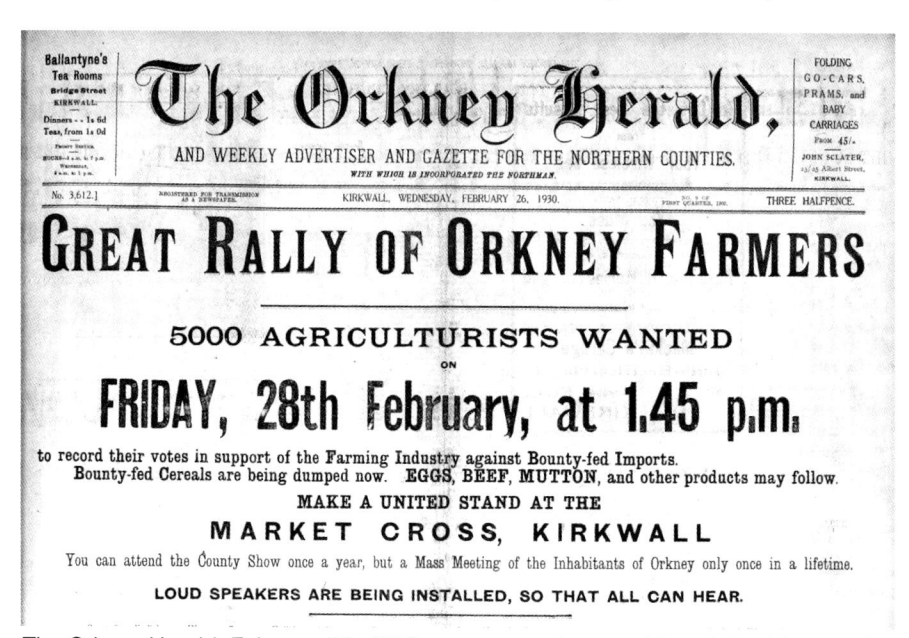

The Orkney Herald,

AND WEEKLY ADVERTISER AND GAZETTE FOR THE NORTHERN COUNTIES.

KIRKWALL, WEDNESDAY, FEBRUARY 26, 1930.

GREAT RALLY OF ORKNEY FARMERS

5000 AGRICULTURISTS WANTED
ON
FRIDAY, 28th February, at 1.45 p.m.

to record their votes in support of the Farming Industry against Bounty-fed Imports. Bounty-fed Cereals are being dumped now. EGGS, BEEF, MUTTON, and other products may follow.

MAKE A UNITED STAND AT THE
MARKET CROSS, KIRKWALL

You can attend the County Show once a year, but a Mass Meeting of the Inhabitants of Orkney only once in a lifetime.

LOUD SPEAKERS ARE BEING INSTALLED, SO THAT ALL CAN HEAR.

The Orkney Herald, February 26, 1930, encouraged support from island farmers to protest over the importation of oats from Germany of landowners, farmers and agricultural employees. This was held in the County Auction Mart, Stranraer, and was organised by T. M. Newbigging, the Union organiser. The unanimous finding of the meeting, according to the minutes, was that the farming industry could only be maintained, and the present workers employed, by the introduction of some method of controlling imports; thereby, securing an economic price for the consumers and farmers for home agricultural produce.

In future years, Stranraer was to be the venue of other Union protest meetings with the action moving to the dock area where demonstrations over the importation of Irish cattle took place.

In Cupar, Fife the local silver band playing "The Farmer's Boy" led the march from the Corn Exchange to the local park. There they heard, through a new fangled contraption called a loudspeaker, the local Union chairman John Arbuckle complaining about the tonnes of oats being imported leaving farmers with losses "of between £500 and £800." He was joined on the bandstand by John Paul, who chaired the local branch of the SFSU. Paul raised a cheer from the workers when he complained about the reduction in wages they were currently getting.

As we shall see later in this chapter, these and other Union organised protests help persuade the UK Government that some controls had to be exercised on imports and equally producers had to have some control over the quantities of milk, cereals, potatoes and livestock being produced. In other words, Marketing Boards and these came into being in 1935.

Income Tax arrives

Ironically, one of the first issues the Union had to deal with was the consequence of farming becoming more profitable in the War years. In the early days of last century, tax was based on half the rent being paid for the farm. But in 1918, the politicians saw an opportunity to try and balance the national books following the horrendously costly war by increasing that demand to 200 percent of the value of the rent.

The tax collectors also saw a large number of farms had moved out of the tenanted sector as estates sold off farms in order to pay off Death Duties for family members lost in the Great War.

There was also a requirement to pay Income Tax. This saw the Union circulate its membership with an explanatory leaflet and specimen cashbook, prepared by the Union's auditor James Allison, advising farmers on the benefits of keeping accounts.

Branch records show Union officers going to meetings to inform unwilling farmers of the new system. The general attitude of the members was that 'their business was their business and none of the Government's business.' A measure of the success or otherwise of this campaign comes in the annual balance sheet of the Union, which in 1920 showed sales of "Book keeping books" bringing in £15 four shillings and a penny (£15.20).

It was ironic that a consequence of the introduction of this fiercely opposed piece of legislation is that, such was the unprofitable nature of farming in the 1920s, that very little income tax was gathered.

Some 20 years later, in the first years of the Second World War, the Union estimated that 25 percent of farmers kept no formal books on their businesses.

Labour problems, pay and conditions

Another of the big issues the Union had to deal with was farm workers' wages. After years of labour shortages during the war, suddenly there were thousands of ex-soldiers looking for work and pressure to reduce pay and conditions built up with this surplus.

In those days, agreements between farmers and their employees were fixed on a one-to-one basis at the local Hiring Fair. These individual deals were, depending on the part of the country, either annual events or six monthly.

To help this system of employment, Union Areas met with representatives of the SFSU in order to set proposed rates but still leaving the deals to individuals. Some Branches also tried to balance those needing men and those looking for work.

Such was the importance of the Hiring Fairs that Perth Area set up an indoor fair

Farmers and their employees engaged on a one-to-one basis at the local hiring fair, pictured here is Dalkeith Hiring Fair, circa 1934

at Little Dunning so that bad weather would not interrupt such a significant event in the farming year. In season, this Fair was open every Saturday evening for two hours to allow men to attend without leaving their work during the day.

But Inverness and other Highland Areas, with support from the SFSU, pushed for the industry to move away from the annual Hiring Fairs with the inevitable annual movement of men from farm to farm. They wanted to see a more permanent workforce.

Education chiefs in the local authorities also backed the change as school routines were broken every May or November whenever the Term Day fell in that particular area.

The SFSU's stance was not entirely altruistic. Under the old system, it had no sooner set up a local branch than it collapsed with its members on the move to another area.

But regardless of the system for hiring or keeping workers, with the collapse in the prices for livestock and grain, the big pressure from farmers was to reduce pay rates and change working conditions.

What were the conditions in those days? Helpfully, Moray Area gave evidence in their report to National Council stating: "Hours of service of farm workers have been agreed at 9 hours per day, not including stable work or mealtimes. The morning yoking is from 6.30am to 11.00am and the afternoon from 1.00pm to 5.30pm. One half holiday is given every calendar month except at harvest time and six days holidays including Christmas and New Year but excluding Hiring Fairs."

An integral part of the deals made at the Hiring Fairs was the bundle of perquisites, or perks, that were on offer. Apart from the tied house, they included the supply of oatmeal, potatoes, milk, and sometimes eggs.

Those near fishing villages could add salted herring to the list. The keeping of up to a dozen hens and/or a pig was also included in some deals. Again thanks are due to Moray County who worked out the value of perks in 1923. On an annual basis, they put the value of the house and garden at £6; the supply of 65 stones (0.4 tonne) of oatmeal at two shillings (10 pence) a stone; three pints of milk a day at one shilling 8 pence (8.3 pence) per gallon; two tons of coal at 45 shillings (£2.25);

From the minute books...

December 1919, Beith Branch

Matthew Gilmour read a letter of complaint with regard to a want of lighting in the early mornings at Barrmill Station. He was then instructed to communicate with the railway superintendent asking him to leave the two lampposts at the loading bank and the lamp at the entrance to the station lit at 6.15am for the convenience and safety of those sending milk by the morning train.

fifteen hundredweights (0.75 tonne) of potatoes at 4 shillings per hundredweight (£3 per annum); cartage of sticks, coals and flittings £3 ten shillings (£3.50), amounting to £34 eighteen shillings (£34.90) per annum which added approximately 30 percent to the cash wage.

Workers from towns employed to lift potatoes could also take home a 'boiling' - but this led to another issue for Haddington Branch in December 1924. They were told of one family in Tranent who had sold something like 30 hundredweight (1.5 tonnes) of potatoes with the goodwill, or otherwise, of the farmer for whom they had been working. The Branch resolution to headquarters asked that the Union might consider whether allowances of potatoes to gatherers should cease entirely.

The fallout on pay and conditions between farmers and the workforce came to a head in East Lothian when farmers tried to rip up the agreement made with the farm servants Union in 1919 on wages and conditions.

Haddington Branch asked the Wages Committee of the Union to insist upon "a substantial reduction of wages." Also that lousing (stopping) time should be at the land's end, and not at the stable door. For completeness, the Branch also wanted working hours at harvest time to be altered so that workers would start an hour later and add that hour to their normal stopping time. In winter, the hours should run from "daylight until dark."

Joe Duncan, the organising secretary of the SFSU, told the workers to hold firm and the matter was on the verge of a strike. It was resolved with the introduction of an arbiter who, in his judgement, allowed some additional hours to be worked but with no reduction in wages.

A different view on farming's role in providing employment was given in 1930 by Union honorary president Sir Harry Hope MP, who said: "With unemployment figures rising to nearly two million and showing no signs of improving, agriculture provides employment for the maximum number of persons being under sound and healthy condition in our rural districts. It provides for the nation the largest possible quantity of foodstuffs produced in accordance with approved and up-to-date methods."

By 1934, there was at least some harmony between farmers and their workforce with Dundonald Branch responding to headquarters' request for opinions on collective bargaining as well as wages and working conditions.

Their response stated that, notwithstanding the depressed state of the industry, there had been very slight, if any, reductions in wages in their district.

Married ploughmen were still being paid around 38 shillings (£1.90) per week with, in addition, free house, oatmeal and potatoes. Relations between employer and employee, or as they put it master and servant, were most harmonious. In the event of collective agreements, which the majority of servants did not appear to want, the present happy relations would become strained, the Branch claimed.

On housing standards, the Union recognised as early as 1919 that something had to be done about the conditions of those who lived in bothies. Although today, there is a romanticised view of this type of accommodation, the actual living conditions were, as the Union recognised, somewhat basic.

Union Honorary president James Gardiner, MP put it more strongly: "We want to abolish the bothy system, which is a disgrace to Scotland."

Another problem brought to headquarters by Dumbarton County in 1920 highlighted the number of travelling people who, in those days, wandered through the countryside picking up odd bits of farm work here and there. They lived in tents on open ground and in season they picked potatoes and fruit and also helped with the grain harvest. Out of season they would eke a living gathering scrap metal, horse hair, wool and rabbit skins, all of which they could sell to merchants in the local towns.

James Lennox, an early supporter of co-operative marketing

Wage disputes may have continued down the one hundred years of the Union, but the local warfare was removed with the establishment of the Scottish Wages Board in 1937, some 13 years after a similar organisation was instituted in England.

Hiring Fairs ceased in the Second World War with working in agriculture being designated a reserved occupation. This meant that workers were not allowed to move from farm to farm as they had done previously; thus bringing more stability into the industry.

The arrival of the Wages Board was not universally welcomed with Dumfries in their report to the Union annual meeting in 1938 commenting that, as from 'time immemorial' farmers had been in the habit of making their own bargains, they did not welcome this restriction.

In 1947, the name was changed to the Scottish Agricultural Wages Board and the original twelve area divisions were merged into one body, which to this day still decides on minimum rates of pay, hours of work, overtime rates, holidays and the value of perks. It is the last of the Wages Boards to survive; all the others having been abolished.

First moves into marketing

Almost from the beginning of the Union's existence, it saw that the future prosperity of farming hinged on improved marketing and it made moves to set up an official Trading Association.

In 1921, Union president James Lennox declared: "We are hopelessly out of date in collecting, classifying and marketing our produce. Why? Because, we are not organised. I am often asked does the Union wish to eliminate the middleman? The

Union does not, but the policy of the Union is to grade out middling men and to abolish middling methods. The middlemen themselves lack organisation.

"We are not suffering from over-production. A hundred million people are starving in Europe with thousands of tons of food rotting here. We are suffering from lack of unity and lack of organisation."

Because the Union was very much a grass roots organisation in its early days, some Branches took the initiative on collective purchasing. For a number of years, a number of branches, including Anstruther, organised the bulk buying of binder twine - a major item of expenditure for arable farmers.

In the mid 1920s, true to its roots, the Union encouraged the co-operative selling of milk through what was called the Scottish Milk Agency.

The idea of collectively selling milk first appeared in the minutes of the Kilmarnock Branch meeting held on 4th February 1924 where members unanimously agreed to the setting up of a milk-selling agency.

A mass meeting was held in Kilmarnock the following month. It was addressed by Union president James Lennox and A. W. Hunter, general secretary.

Shortly thereafter, Dumbarton Branch reported: "the supply of milk to private retailers is sold through a Committee of Producers and the price collected by the Branch Secretary, continues to give unqualified satisfaction. The arrangement prevents undercutting, gives equality of treatment and secures prompt payment. The scheme is worthy of wider application and it seems to the Committee that only by the adoption of some such method can a solution of the recurring milk difficulties be found."

Two years later, Dumfries County reported to the National Council: "The most important item, so far as dairy farmers are concerned, was the establishment of the Milk Selling Agency in which your Union took an active part. It is expected that the Agency will be the means of stabilising and regulating prices and that the hawking of surplus milk, which in the past was responsible for rendering dairy farming an unprofitable occupation, will be a thing of the past."

In the foreword to the Union's annual report that year, Alexander Murdoch, chairman of the Scottish Milk Agency wrote: "This organisation has now been in existence for 18 months and during that short period it has gone from strength to strength, not only as regards membership but also as regards usefulness."

From the minute books...

1925, Nairn County
Nairn County wrote to Headquarters asking for a tax to be imposed on all foreign barley delivered at distilleries, maltsters and breweries. If such were done, it was hoped that there would be a demand for home barley.

As at 1st November 1927, its membership was 1,201 and its turnover in milk was 19 million gallons per annum. A year later its membership had risen to 2,478 and its turnover in milk to 24 million gallons per annum.

However, with a wider collapse in the value of milk, the Agency went out of business shortly thereafter.

In a similar move, triggered by low prices of wool, the Union was instrumental in establishing the Scottish Wool Growers' Society. In 1928, Aberdeen County expressed its "gratification at the steady and excellent progress being made by the Scottish Wool Growers' Society."

This was, said the Aberdonians, an example of farmers setting up an organisation for the marketing of a commodity, which they have to dispose of. "The value which the Society can be to its members is in direct proportion to the support accorded to it."

That same year, Cassels Jack, chairman of Scottish Wool Growers was quoted as saying: "The net profit earned on share capital for the second year of trading is equal to 18.45 percent, and producers would do well to ponder whether they prefer to secure that profit for themselves by joining the Scottish Wool Growers or would they allow the broker to have the profit on the sale of their wool." Cassels Jack, who would four year later become president of the Union, emphasised the word "their" in his comment.

By 1930, Union members were being encouraged to give "every possible support to the Co-operative Marketing Enterprises, which have been introduced and which are being fostered by the SAOS and the Union."

Scottish Wool Growers advert in the Union's Annual Report of 1929

Marketing Boards; their arrival

The 1931 Agricultural Act, which had been promoted by the then Under Secretary of State Tom Johnston MP, was not considered sufficiently robust by those farmers keen to protect and promote their own produce.

It was later strengthened with the 1933 Marketing Act, which was brought in by

Minister of Agriculture and Borders farmer, Walter Elliot MP. He also managed to get quotas on imports during his term of office; a move welcomed in Berwick's Annual Report. They described the import quota scheme for beef, mutton and pork as "already having an effect on prices."

Underlining the status of the Union by this time, the Scottish Secretary of State, Sir Godfrey Collins was invited to address the 1933 annual meeting; the first time the top politician in Scotland had been given the honour.

He declared his predecessors had lost much through not being at the meeting. "Do not think for one minute that your criticism has ever been impatient or ill-informed. I do think it has been made in the right spirit."

Then in a show of sympathy with the difficulties faced by farmers at the time, he added: "I am a manufacturer and I am glad to think I have not to sell my products at pre-war prices."

Walter Elliot MP, the man behind the 1933 Marketing Act

By 1935, the Marketing Act saw the introduction of marketing boards for several commodities including milk, potatoes and pigs. Others were planned for raspberries, eggs, poultry, milk products and sugar beet.

Renfrew Area summed up the new legislation by saying it was increasingly important for farmers to be members of the Farmers Union. "A complete transformation has taken place in the agricultural world. The long settled fiscal policy of the country has been changed, and quotas, control of imports, tariffs and fixed prices have become part of the economics of agriculture. Alongside these changes and related to them, there have been set up Marketing Schemes for milk, pigs and potatoes with eggs and poultry soon to be added. These great changes, which it is hoped will be for the good of the farmer and the community, could only be initiated, criticised and controlled by the existence of such a body as the Farmers Union."

From the minute books...

1928, Hawick Branch

The Branch raised the issue of closing the High Street for "sheep traffic." They received permission to proceed along the street until 9 o'clock.
In the same year, the Branch also arranged for the repair of a drain on the drove road at Calaburn as this road was greatly used in driving sheep to the sales and the County Council did not keep it up.
It was left for Hawick Branch and Selkirkshire County to go fifty-fifty. The account came to £3 but Selkirkshire only gave £1. The Branch minutes stated, "We just paid the rest."

The three Milk Marketing Boards in Scotland - the Scottish, the Aberdeen and the North of Scotland - were not universally welcomed with Union Branches in the East of the country complaining about powers to take surplus summer milk off the market in order to make butter and cheese. With a smaller production of milk in the East, most of the milk could be sold more profitably on the retail market.

The milk board scheme was viewed as helping West of Scotland dairy farmers as their production was largely based on summer production.

However, once this initial disagreement settled down, the three milk boards monopolised the Scottish milk market right up to 1994 when the UK Government stripped the boards of their powers.

In 1936, describing the new Boards, Bob Boothby MP for Aberdeen & Kincardine East, when addressing the annual meeting said: "I think that on the whole the Potato Marketing Scheme is working; that on the whole the Milk Scheme is being mended; and that the Pig Scheme, as at present organised, should be ended."

In a further comment on milk marketing, Boothby said: "The rights and interests of the middlemen must be protected but they are not sacrosanct. I still believe that the gap between the price paid to the producer and the price paid over the counter is far too wide."

Not everyone was happy with Berwick branch, which had supported the import quotas, describing Marketing Schemes as 'a complete failure.' Thornton Branch reported their members were losing confidence in the Marketing Boards and claimed the deal by the Government with promises with regard to imports being balanced by producers supporting Marketing Schemes was not working. "We are still looking for the right to have the first call on our home markets," they minuted.

Recognition of the Union

By 1921, the incoming president James Lennox was able to comment that in less than eight years since the Union was instituted: "We have now 143 branches and about 50 percent of the farmers of Scotland as members.

"Early in our existence the country was plunged into a great war when our industry like others was subject to Government control, and the Union had to fight against many vexatious and sometimes unnecessary restrictions. "But while claiming and generally securing reasonable justice for our members, aye, and non members, I speak with knowledge that no Union of employers in the country, with the same strength and influence, took less advantage of the nation's necessity or rendered more assistance to the Government."

In 1922, Robert Greig, the secretary of the Board of Agriculture for Scotland - the forerunner to the Department of Agriculture - said of the Scottish Farmers' Union: "With regard to the questions of agricultural policy which arise almost daily, the advantage to the Board of Agriculture for Scotland and to the Secretary of State for Scotland of obtaining the opinion from such a representative body as the Scottish Farmers' Union cannot be overestimated."

He then made his Government's impartiality known by adding: "So long as it is recognised that the Board of Agriculture for Scotland is not the advocate of one class, but is concerned with the promotion of agriculture in the widest sense, in accordance with the policy of the government of the day, nothing but good can come from the close association of the Board and the Union."

The depressed state of farming seeped through to the Union when the annual meeting in 1930 faced a proposal from Fife Area calling for a reduction in the subscription rate, which was then based on farm rents. The proposal to reduce the subscription from one penny (0.4 pence) to a halfpenny (0.2 pence) in the pound was rejected.

Captain Alex Manson who masterminded the link-up

A few years earlier, Moray County believed the benefits from Union membership were not sufficiently known locally. "There is an impression, not without substance, that the resolutions of the Union have not been loyally adhered to."

In 1939, it was announced that the Union and the Chamber of Agriculture for Scotland, which had been in existence since the previous century, had officially amalgamated. President at the first annual meeting of the clumsily named "Scottish Farming Union and the Chamber of Agriculture for Scotland" was Captain Alex Manson who remarked it was better to have one strong organisation rather than two. The link up was successful with the membership of the new body rising to more than 18,000 by the end of 1945.

Within a decade, the shorter name of the National Farmers' Union of Scotland was adopted.

From the minute books...

1928, Dundonald Branch
A claim was made by David Goldie, Barassie, Troon for the loss of two yearling Clydesdales which were run over and killed by a motor bus owned by the Midland Bus Company. After full consideration it was unanimously agreed to recommend that the County Executive Committee give what legal assistance was necessary in the prosecution of the claim.

From the minute books...

1922, Potato trading
One of the big issues for the Union was the great danger of distributing Wart Disease in potatoes from Continental imports. The Union urged the Government to stop those imports but twelve months later Union vice-president Alex Batchelor, a Dundee based potato merchant, was urging the same Government to consider exporting surplus potatoes to Continental countries.

Pest Control
In one of the more difficult to understand minutes, in 1922 Selkirk Branch thanked the members of the Crow Committee for the energetic manner in which they carried out their duties. No explanation was provided as to what these duties might have been.

1923, Early Union Victory
Selkirk County reported the Union had successfully negotiated livestock rates with the railway companies. These included "half rates for the return journey of wintering hoggs, with one extra wagon free for every four on the outward journey."

National Insurance
One of the more controversial pieces of legislation in the early years of the Union was the introduction of National Health Insurance deductions from workers' wages. In 1926, Berwick County reported it had endeavoured to get the rural workers' contribution of NHI reduced, as statistics proved that there is less sickness amongst rural workers than industrial workers. Berwick also protested against agricultural workers being brought under the Unemployment Insurance Scheme.

1930, Pest Problem
At a conference in Aberdeen between the NFU and the Land and Property Federation, the damage caused to farm crops by rooks was discussed. Colonel Duff, of Haddo, said that, since fences had taken the place of hedges, crows were sitting on fences watching the nesting. The crows then they harried the partridges' nests and he claimed that was the cause of a scarcity of partridges in Scotland.

1933, Road Tax
Lanark Branch put forward a resolution taking strong exception to the increase of taxation on motor vans and lorries. The Branch considered that road transport already paid in horse power tax and petrol tax an amount quite sufficient to cover the whole of the upkeep on roads. The resolution added "the motor lorry has become so useful to farmers that no obstacle should be placed in the way of its extended use."

The Second World War Years

After surviving the economic depression in farming that dominated the 1920s and 1930s, farmers were once again asked to feed the nation in time of war

At the start of the conflict, using rhetoric similar to Winston Churchill, the Union president William Wright from The Heugh, North Berwick called on the industry to roll up its sleeves and produce food.

"We are engaged at this moment in what is likely to be the greatest struggle that has ever taken place in the history of the world. Lined up against us are the vastest armed forces ever known. It is a war not only of armies but a war of nations in which every individual must play an important and vital part. If that is true of the people of the country it is more than ever true of the agricultural industry, which can be described as the fourth line of defence.

William Wright encouraged war time production

"I should like all farmers and the public generally to realise that the work of the Union can only be designed to secure that measure of return which is necessary to allow the industry to do its duty at this time, and I would personally deprecate any idea that farmers should profit at the expense of the country at this time of national emergency." His oratory produced shouts of 'Hear, Hear' from the delegates fired up with national pride.

Yet in the run up to the conflict, the Union had been in angry disagreement with Government over the continued high level of imports. In July 1939, Prime Minister Neville Chamberlain had addressed a farmers' meeting where he had stressed the need to import food in order to gain export orders for machinery made in the UK.

The Prime Minster claimed the Government's agricultural policy had been highly successful. Both statements caused the Council of the Union to be profoundly dissatisfied according to the minute taker.

The meeting scribe continued that "no time was lost in acquainting the Prime Minister of the deep concern and disquiet which his statement had created amongst

Scottish producers. A reasoned memorandum on the conditions prevailing in the industry at the time was forwarded to Chamberlain and the other ministers concerned, as well as to all Scottish MPs."

Another pre-war example of the attitude of Government to home food production emerged when the Union Council appealed to the Government to assist the livestock industry by arranging for the supply of home-produced meat to His Majesty's Forces.

The War Office response to the Union stated: "considerations of price and of administrative convenience make it essential for the Army Council to rely in the main on supplies of frozen meat from the Dominions. Accordingly we cannot accede to the demand that home-killed meat be supplied in preference to imported meat."

Within months, these supplies of imported meat were cut off and the Armed Forces relied heavily on home produced meat for the rest of the war.

Other examples of the tide of imports hitting farming in the pre-war era include the Union having to deal with concerns from poultry producers about the serious state of the industry caused by eggs coming in from abroad, particularly from China.

Sheep producers also made a series of complaints to the Union's Livestock committee and the Council over low prices for fat sheep as a result of imports coming in from the Dominions.

And with the Second World War clouds on the horizon, at the 1939 annual meeting, A G Norrie of Cairnhill, Turriff expressed concern over the large amount of foodstuffs being imported into this country.

He noted there was a new UK Minister of Agriculture, Sir Reginald Dorman-Smith and as Dorman-Smith was a former English NFU president he believed some positive action to benefit farmers should result. However, Dorman-Smith was moved from his Ministerial position at the start of the Second World War before having time to influence anything.

Norrie, in a poetical vein, then claimed: "Millions of acres of land (in this country) are weeping their eyes out for the plough."

Within months and with war on the doorstep, he and thousands of other farmers had their wish as the UK Government called for an additional two million acres to go under the plough as part of their Home Food Production Campaign.

For their part, Scottish farmers increased their cropped acreage by some eight percent in the first year of the conflict, growing an additional 250,000 acres of food.

Their reward for doing so came a year later at the next annual meeting of the Union when Tom Johnston, Secretary of State called for another 250,000 acres of land to be ploughed up and put into crop.

By 1943, Johnston was still pressing for more food production. "During the past three years the tillage area in Scotland has increased by no less than 41 percent, and for the added foodstuffs so necessary we must birse yont deeper onto (go further into) the marginal land."

To encourage the ploughing of this more marginal land, he announced a subsidy scheme for ploughing up.

The arrival of war may have increased the importance of home food production but it decreased the role of the Union because the Government had put in place local Agricultural Executive Committees (AEC) with wide ranging powers.

The AECs' remit ranged from persuading farmers to plough up more grassland through to deciding which of the farm workforce would remain on the farm in a reserved occupation and which individuals should be sent off to fight. The AEC could and did take over badly run farms; a role they carried out until well into the 1950s.

Almost all of the marketing of farm produce moved out of private hands and into the control of Government. In January 1940, the Government took over responsibility for the purchase of fatstock at collecting centres across the country.

With milk, the Government recognised the paramount importance of an adequate supply so it assumed responsibility for milk policy including production, price and distribution. In the first year of the war, price increases of between 2d and 3d (approx. 1 penny) per gallon were announced to try and achieve sufficient supplies.

The control of the sale of cereals was most contentious with growers only being allowed to sell to approved buyers. In 1940, the Government decided to restrict the output of whisky and other potable spirits by one third of the production of 1939.

This restriction was intended to increase the tonnage of cereals available for animal feeding but this caused consternation amongst growers who protested to the Union that government had removed two thirds of their market. They wanted some measure of compensation for the loss of this market.

Because the Government valued poultry - both for their efficient conversion of feed and also for their egg laying ability - one third of the supply of home grown wheat was earmarked for stock feeding but this did not prove to be sufficient, so further allocations were made available.

From the minute books...

1941, Papermaking
The War Committee of the Union agreed a price of £4 per ton for baled straw on the farm for papermaking. Farmers were told papermakers would take all the straw they could because of their inability to get supplies of esparto grass.

Bits and bobs
That same year saw a number of unusual items on the War Committee agenda. These included the dilution of Cod Liver Oil, the prices given for rabbits, the extermination of rats, a Licence Duty on dogs and National Air Raid Precautions for animals.

Within weeks of the start of the war, the newly linked up National Farmers' Union and Chamber of Agriculture for Scotland decided to abandon all commodity committee meetings leaving only a War Committee composed of the Honorary President, President, Vice President and the Conveners of the Standing Committees.

Among the early issues dealt with by the War Committee was the shortage of machinery to tackle the increased acreage of cropping. Orders for tractors were taking months to fulfil as factories turned out tanks and military equipment as a priority.

Only later in the conflict did the 'lend lease' scheme with the United States bring in thousands of tractors and hundreds of combines and other essential farm equipment.

The Union also acted on behalf of a number of members whose land had been taken over by the military authorities, the Admiralty or the Air Council for defence purposes.

By 1942, the restricted amount of animal feed caused the Union War Committee to consider that beef herds in Scotland might have to be culled. However, the Government introduced a subsidy of £4 per head for cattle in Scotland's hill ground. It was initially limited to the Highland and Galloway breeds, and crosses of these breeds with a Shorthorn bull but this restriction was removed within the year.

The severe winters of 1940 and 1941 prompted the War Committee to ask for an increased hill sheep subsidy as the Committee considered the 2/6 (12.5 pence per head) inadequate.

There were also pricing battles on potatoes where the War Committee succeeded in gaining a 5/- per tonne (25 pence) increase for Kerrs Pinks and Redskins, and in sugar beet where the £4 per ton being offered was considered inadequate.

With virtually no imports of fruit during the war, pressure was placed on growers in this country to produce more. But storm damage in 1943 and frost problems a year later badly hit raspberry production. The Union successfully pushed for increased contract prices for fruit.

Summer time and long-term problems

During the Second World War, the Government introduced a scheme advancing the clocks by one hour ahead of Greenwich Mean Time (GMT) in the winter and two hours ahead of GMT in the summer but the Union did not welcome this arrival of Double Summer Time (DST).

Union President William Graham, speaking at the 1942 AGM said: "Undoubtedly the enforcement of Double Summer Time a year ago had the effect of restricting production not because farmers did not want to work to the very limit of the hours of daylight available but simply that Scottish conditions precluded them from

William Graham CBE railed against changing the clock

doing so.

"I hope that as a result of a resolution which was practically unanimous, it will be possible to influence the Government to relate Summer Time as applied to agriculture to the previous extra hour only."

But, Scottish Secretary of State Tom Johnston speaking to the annual meeting was not for turning saying that there were "security reasons" on which he could not elaborate on DST being introduced.

Just how farmers felt about DST emerged in a letter to The Scottish Farmer in 1943. "Getting up at 4.30am which is nearly 2.30am by the sun, it is quite dark making it much more difficult to get the cows in and by the time the hens are in at night it is 11pm making a very long day."

However, a year later in 1944 Double Summer Time was still operating causing one area president to complain at the annual meeting. "When it was announced that DST would extend from 2nd April to 13th August it was discussed in many farmers' meetings up and down the country and was denounced in strong language with no dissenters.

"Before men can fight they require food and also before planes, ships and guns can be constructed, they need food. Yet they (the Government) impose one of the most irritating and obnoxious measures on the people who produce food."

The Government argued the daylight saving measures were maintained to allow an extra hour of daylight at the end of the day so that, for example, goods trains could be loaded in unlit sidings.

This cut no ice with one irate letter writer to The Scottish Farmer who commented on the Home Secretary, Herbert Morrison 's extension of DST by 5 weeks: "Mr Hitler must be a very accommodating man when he considers Mr Morrison's DST just to suit him before he sends over his flying bombs."

In 1948, the first post war Prime Minister, Clement Atlee assured farmers that DST would not be re-introduced. This was reported by the Union to have given the "utmost satisfaction to farmers in Scotland" as, unlike in England where most farmers were not excited about the change, the majority of Union Areas in Scotland voted against any alteration to Summer time either at the beginning or end of the period.

The whole issue of adjusting the clocks on an annual basis - moving them forward by one hour in spring and backward by the same length of time in the autumn - started in 1916 but even in the early days, British Summer Time (BST) was not a popular move with farmers.

In 1920, the County of Dumbarton protested to their local MPs, the Secretary of State and even the Prime Minister, when it was made clear that BST would continue to operate.

That same year, at the annual meeting of the Union, Mr T H Ryland of the English Farmers' Union, moved that Scotland should protest against the continuation of summer time. He stated the whole principle underlying the continuance of this rule was that it was going to be a great convenience to many people to play golf and tennis. Was it fair that farming should be subordinated to a question of recreation he asked?

While the unloved DST was seen off at the end of the war, the Union has had to fight on the 'clock changing' front a number of times since then.

In 1960, the majority of Areas of the Scottish Union voted against any changes being made. This put them at odds with the English Union, which was by now in favour of extending British Summer Time.

Then in 1968, the Government decided to introduce British Standard Time (BStT), which involved advancing the clocks one hour ahead of GMT for the whole year instead of for summer only.

This was a time when Britain was negotiating membership of the European Community and may have been an attempt to appear 'communitaire' or to appease business interests in the south. But it was most unpopular in rural Scotland.

From the outset, the Union opposed its introduction and the Government recognising the strength of feeling in Scotland announced that the new time would be introduced for a trial period of three years.

In an early attempt to get devolved powers, during the committee stage of this Bill an amendment to exclude Scotland from its provisions was only narrowly defeated.

As a side effect of BStT, the Union pressed for an immediate relaxation in the mileage limit for school buses in rural areas stating as the reason: "this winter many children will have to walk to school in darkness, especially in the northern counties."

In 1969, Union Assistant Secretary John Lefley reported on information received from nearly all Union Areas on the effects of BStT during the past winter. The Areas who replied to the survey represented over 96 percent of the Union's membership and all but three of them expressed complete opposition to BStT.

A major concern expressed was that farm employees were very angry about the danger to their children going to school on dark mornings and many respondents threatened to seek alternative employment in or near towns if the experiment continued. It was also stated that it was easier to work into darkness than start the

From the minute books…

1941, Bracken Control

Tom Johnston, Secretary of State, in his speech to the annual meeting said: "I have been interested in an experiment by Lord Trent at Ardnamurchan where an effort has been made to feed pigs in the open air upon the rhizomes or roots of bracken. "They manure the ground as they go along and Lord Trent has informed me that he is, this year, ploughing land that has not been ploughed for a century. The difficulty, however, appears to be that the pigs require some meal along with their mashed potatoes and bracken.

day in darkness.

A year later, a Department of Agriculture survey into the effect of BStT on farming showed 78% of farmers had been adversely affected in some way.

Fraser Evans, convener of the Union's Labour committee, referring to a report on the three-year trial of BStT said: "It is absolutely clear now that this foolish experiment must be ended. The nebulous advantage to London businessmen of having longer hours to phone the Continent is obviously outweighed by the crippling extra hours and inconvenience caused to agriculture and other vital industries, particularly in Scotland."

And that was that until 1990 when Robert Lamont, convener of the Legal & Commercial Committee, appealed to all Scotland's MPs from every party and Scottish Lords as well, to throw out the notion of putting Britain on European Standard Time. "People in the south have little perception of the scale of the inconvenience this idea would cause. Scotland has less daylight in winter in any case – and the fact that Scotland lies further west as well makes European Standard Time doubly inappropriate."

There may be more stock under cover in winter nowadays and farm buildings and machinery are better lit than they used to be but there are of necessity many jobs which need natural daylight, and the arguments that Robert Lamont made in 1990 are still relevant today.

A new crop arrives

Right up until the First World War, the UK had depended on fulfilling its need for sugar by importing it from Commonwealth countries where sugar cane plantations were established. This was big business and fortunes were made both in the growing and in the trading of sugar to the extent that it was called 'white gold.' However, the fragility of this trade was exposed once enemy warships started sinking cargoes of sugar causing a shortage in this country.

In the 1920s, the Government, having been caught short during the war through an almost complete cessation of sugar imports, encouraged the growing of sugar beet through the massive use of subsidies for processing factories.

This saw a proposal for a refining factory being built at Cupar in Fife but first there was a need to prove the crop could be grown in Scotland. The Union was heavily involved in establishing field trials with the first results ending up with yields of well under 10 tons (tonnes) per acre and the crops having a closer resemblance to carrots than to well grown beets.

William Bruce improved terms for beet growers

However by 1926, in anticipation of the opening of the sugar beet factory, East Lothian County of the Union reported: "Except in one or two cases, the growing of sugar beet has been a success. The general treatment of the land on the whole has been the same as is given to potatoes or turnips, except that a little more attention has been given to the depth of cultivation."

By 1935, president William Bruce told the annual meeting that after prolonged negotiations with the owners, the following terms were agreed – 37 shillings (£1.87) per ton (tonne) on a 15.5 percent sugar basis, free on rail, with an additional sixpence (2.4 pence) if the quantity grown exceeded 6,000 acres.

Thinning beet at The Grange, Kirkcaldy, circa 1957

These terms were received by growers with such goodwill that, although late in starting to make contracts, 8,240 acres were contracted. The average crop was 8.46 tons (tonnes) per acre.

But in 1939, the annual meeting of the Union was told that little more than 8,000 acres were grown in Scotland, which was some way short of the 10,000-acre target. Union members were warned that unless the target acreage was reached, the cost per ton for processing in Scotland would continue to exceed that for beet processed in England.

Throughout the Second World War, there was a constant battle between the AECs, which were pushing for more production of the crop, and farmers who either claimed growing sugar beet was unprofitable or that their farms were unsuitable for growing it.

By 1945, the Union's Sugar Beet Committee felt so strongly that many farmers were being forced to grow sugar beet at a loss that, at a meeting on 17th January, a resolution was unanimously adopted. This called for the Department of Agriculture to withdraw cultivation directions for this crop in the event of continued failure by the Government to offer an economic price.

In putting this resolution to the Department, it was made clear that there was no suggestion of strike action. The intention was rather to impress upon the Government the serious dissatisfaction felt by farmers at a policy which insisted on growing beet on land unlikely to yield a tonnage which, at a price of 81 shillings (£4.05) per ton was insufficient to meet the expenses involved.

The matter was resolved but in the post war years, the acreage of sugar beet grown fluctuated wildly. A new record acreage of 14,400 acres was grown in 1955 and three years later the Union heard the valuation of that year's crop had exceeded £1 million.

By 1969, Andrew Arbuckle, convener of the Union's Sugar beet committee announced that the British Sugar Corporation had decided to close the Cupar factory at the end of the 1971 season.

The Union fought a long and hard campaign to keep it open but the last loads of beet were processed in December 1971.

Union vice president at that time, Jim Stobo, recalls one ploy by the campaigners to help keep the factory open. This was an invite to a top-level civil servant to a day's shooting in the Borders. Despite getting all the best pegs (stands); the civil servant did not hit a bird. No one will ever know how important these misses might have been.

A shortage of Labour

Almost from the beginning of the hostilities in 1939, the main Union battle was waged on the balance between producing more food but with fewer men on the farms. For example, in the spring of 1941 the Government said that Scottish agriculture must contribute a quota of 10,000 men to the fighting forces. The Union received this demand with some consternation. Its position was that the labour shortage had already assumed serious dimensions and the success of the 'ploughing up' campaign meant that a corresponding accumulation of work had to be undertaken on the farms.

The position was eased when the Minister of Agriculture said that the men selected would not be taken away from the farms until after harvest and, in fact, they were not taken away to fight until the turn of the year.

One of the problems faced early in the years of conflict by the Union was dealing with the occupation of tied houses by the dependants of farm workers who had been called up to serve with the Forces. Working with the Scottish Farm Servants' Union (SFSU), the War Committee was able to bring about a settlement in most of the cases reported to them.

But the combination of increased acreages of crops to be harvested and reduced numbers of men to deal with the workload was a constant theme on the agenda of the War Committee.

It saw the Union support a number of initiatives aimed at getting more people to bring the harvest home. These included the release of soldiers on a temporary basis to help with the harvest.

The Women's Land Army (WLA) proved its worth throughout both World Wars, pictured here at Eaglescairnie Farm. The girls worked the land while the UK's labour force was at war. The Land Girls, as they were known, were realeased from their duties, when the WLA was disbanded in 1951

On the other hand, large numbers of farmers and their workers were among the first voluntary members of the Home Guard. The Union was asked to take up the cases of those volunteers who were being prosecuted because they found it impossible to attend parades at busy periods.

The Union pointed out that certain industrial workers, such as miners and shipyard employees, who worked more than 60 hours per week, were exempt from Home Guard duty and it was hoped that this could be extended to those working in agriculture who regularly worked these hours at busy times.

In another attempt to tackle the labour shortage, city dwellers were encouraged to come to the country to help in the picking of fruit and potatoes as a 'holiday' away from the trials and tribulation of urban life during wartime.

The Women's Land Army (WLA) came into its own and, after initial scepticism about the abilities of this new workforce, it proved their worth with the War Committee soon asking Government for more Land Army girls. Initially the WLA mostly worked as milkmaids or they looked after the hens but soon they were driving tractors and working other machinery. A special section of the WLA was set up to control pests such as rats and mice; equivalent to the modern day pest control companies.

In the latter stages of the war, use was made of the significant numbers of prisoners

Prisoners of War, who were used as farm labour during peak periods. Pictured here at Newburgh, Fife, in 1943

of war at peak labour periods, but this was not without its own issues. In general terms, Union members wanted Italian POWs for their attitude but German ones for their work ethic.

Refugees from war torn Europe also arrived in Britain as the war progressed and many of those Displaced Persons found work on farms but one source of labour was cut off in 1945. Because of the Irish political situation, restrictions on travel from Eire were imposed. This seriously reduced the number of Irish labourers available for the potato, grain and sugar beet harvests.

On this issue, the Union recorded: "It is hoped that normal numbers will at least be available this year and that a system of control will be introduced in order that full advantage may be got from this source of labour."

The arrival of Displaced Persons onto farms brought its own problems with the Union having to deal with a situation whereby a charge of two shillings (10 pence) per hour was made in respect of DPs who did not understand English and could not work without constant supervision, although a charge of only one shilling eight pence (8 pence) per hour was made in respect of German prisoners. It was suggested that the rate in future for DPs should be the same as for the rest.

In contrast with the ending of the First World War, where large numbers of men returned looking for work on farms, the concerns of the agricultural industry in the 1940s centred on how agriculture would cope with labour shortages when hostilities ceased.

As the war drew to a close, the Union recorded: "Agriculture may be deprived of much of the auxiliary labour now available and it is essential that the labour force is maintained to meet the high level of production which will still be necessary."

If anything these concerns were magnified when in 1948, the Scottish Agricultural

Wages Board decided that stock and dairy workers would be free from Saturday mid-day until Monday morning fortnightly during summertime and once in four weeks in winter hours.

Non-stock workers' week would be reduced to 47 hours with the addition of 5 and 7 hours respectively for tractor and horsemen.

Another factor affecting labour supply on farms was the lack of housing and the Union called on local authorities to build more properties in rural areas for farm workers. This call was successful with a number of projects being undertaken.

The conclusion of hostilities

The fighting may have stopped but the big concerns in the farming community, and therefore the Union, were the shortages of inputs such as feeding and fertilisers as well as fuel for getting produce to markets.

The Union annual report in 1946 commented: "the fertiliser supply position has continued to be difficult and supplies of potash have failed to come up to expectations. Spare parts for tractors and implements are now coming forward in better supply."

A year later both the pig and poultry sectors, which had seen feed supplies cut during the war, complained to the Union that the reductions in rations of essential foodstuffs had still not been lifted. But it would be 1953 before feed allocations to pig and poultry producers were lifted. In the case of pigs, the feed allocations were based on the amount of meat delivered to the bacon factory. For every 70 kilos of pig meat produced, the farmer was supplied with 150 kilos of feed.

Egg producers also received feed based on their output. Here the figure was 70 kilos of poultry feed for every 80 dozen eggs sent to the packing stations.

Shortages, food rationing and controls brought their own issues and, in 1947, the convener of the Poultry committee attended a meeting in London at which it was announced that the Ministry had decided to take steps with a view to curbing the illegal trade or black market in poultry. In future, only licensed purchasers would be permitted to buy stock birds and the prices of table poultry and turkeys would be increased.

Another shortage to hit the farming industry was transport. In 1947, potato growers in Aberdeen Area experienced considerable difficulty in arranging the shipment of seed to England. The shortage of shipping space was attributed to the return to their own countries of foreign vessels, the use of which had been available in Great Britain during the war years.

One crop that disappeared from Scotland following the war, despite efforts by the Union to retain it, was flax. During the war, flax fibre was used to make parachutes, tents and camouflage but with the conclusion of hostilities the demand for flax fell. The Ministry of Supply intimated that the factories at Turriff, Blairgowrie and Cupar would be closed. Any flax grown in Scotland would therefore require to be grown for and consigned to such factories as were operating in England.

By 1946, the new Secretary of State for Scotland, Joseph Westwood MP was looking to a more certain future for farming. He declared: "The war against our enemies has finished but the war against famine in large parts of the world goes on."

He also announced his Government would establish as "an essential and permanent policy" for food and agriculture, a system of assured markets and guaranteed prices for the principal agricultural products, namely milk, fat livestock, eggs, cereals, potatoes and sugar beet.

This heralded the introduction of the Annual Price Review with the Union playing an integral part in determining commodity prices.

Looking to the future, the Union established a General Purposes committee in 1946 to deal with matters outwith the remit of the existing standing committees.

For the first time, the quality of produce came onto the agenda of the Union and it did so in a number of ways. Following complaints about Scottish seed from English growers, the Department of Agriculture was recommended to set up an inspectorate to examine seed potatoes on the farm and in transit to check dressing and quality and to take action where necessary.

In 1946, the Union attempted to secure more encouragement for the production of quality beef and mutton, but the continuing need of the Ministry of Food for quantity rather than quality made it impossible to obtain the quality incentive that the Union desired.

In the pig sector, the Union supported a scheme of boar licensing on the basis of that already in place for bulls. The Union's view was that the severe culling of stocks imposed on producers during the war years meant only the better sows had been kept so it was logical to improve the stock through better boars.

Cereal growers fearful of a market collapse, similar to that witnessed in the wake of the First World War, asked for a marketing scheme for all cereals following the Ministry of Food relinquishing control of the sector.

And the Milk committee adopted a resolution calling upon the Scottish Milk Marketing Boards to undertake the wholesale and retail distribution in the interests of efficiency and economy.

And the last word on the war years

In an unspoken truth, farming prospered in the difficult war years. It may be the only time in its one hundred year history, but in 1941 the Union asked Joseph Allison, their auditor for guidance on the liability of farmers for Excess Profits Tax in order to assist members desiring clarification of their position.

Two years later, the Union recorded that with the introduction of Schedule D (profits from a business) Income Tax "many farmers are faced with the necessity of keeping books."

The Price Review Years

Down on the farms in post war Britain big changes were happening. One of the main reasons the farming industry had been able to increase production massively during the war years was because of mechanisation. The horseman may have been an essential part of the pre-war farm but, by the post-war years, tractors were dominant.

However, the Union in 1949 still had to deal with concerns over this wholesale shift towards mechanical horsepower. The Ministry of Agriculture expressed concerns over the decrease in the working horse population of Great Britain.

The Union contended this was because there was a steadily decreasing demand for working horses. Back came the Ministry stating the need for horses might become urgent if circumstances arose which curtailed supplies of fuel for tractors.

The Ministry missive concluded: "In such an emergency the country may well have to revert to horse traction. The national horse population may therefore require to be stimulated as a defence measure."

The official record of the Department of Agriculture seven years later in 1956 revealed that seven percent of the farm workforce was still categorised as horsemen. The same report recorded nine percent of employees as dairy stockmen and 32 percent employed as tractormen.

Apart from this minor diversion, even although it was weary from five years of producing ever more food for the nation, the post Second World War farming industry entered peacetime with more optimism than at any other period in the past century.

The annual report of the Union in 1948 talks of "a new appreciation of the value of a sound home agricultural industry" and commented that every £1 of imported food could now only be acquired at the expense of some other article of living.

"For the home farmer, who has so often seen his best efforts prejudiced by the existence of food surpluses on the world market, the transformation brings high opportunities of service to the nation. Upon him has fallen the task of helping to reduce expenditure on imports and of making a major contribution towards the achievement of long-term national stability."

This should not be seen as unbounded optimism as the industry was bedevilled with shortages and rationing. Coupons were needed to buy food and fuel; an issue that remained well into the 1950s.

The rationing of animal feedstuffs was complex with allocations for the numbers and types of livestock kept. Pedigree stock was allowed more than commercial animals, autumn calvers more than their spring counterparts and rations for horses varied with their working capacity. There were even variations in the quantity of feed allowed for sheep dogs depending on whether they were working difficult hill ground or on easier terrain.

For the first time ever, the farming industry was entering an era where food production was supported. The principal driver for that support was the introduction of the Annual Price Review, which provided the security of markets and prices for almost all farm produce.

The architect of the Price Review was UK Minister of Agriculture, Tom Williams and in drawing it up; he inserted a clause stating the annual negotiations should be carried out with "such bodies who appear to represent the interests of producers in the agricultural industry."

The Scottish Union along with its counterparts in Ulster and England were now at the centre of UK agricultural policy.

The saying 'be careful for what you wish for' might have been coined for the position the Union now found itself in with members wondering why commodity prices were not higher than the negotiators had managed to achieve and also why one sector seemed to be better rewarded than another; the complainants changing on a yearly basis, depending on the outcome of the annual deal.

The early Price Reviews were carried out against a background of a world recovering from the worst conflict ever known. The economic reverberations arising from a move back into peacetime often caught the economists and negotiators short. Supplementary price reviews were held to sort out late changes or shifts in emphasis and inflationary effects on costs and prices; all of those problems adding further to the burden of the negotiators.

The report on the 1950 Price Review commented that: "the atmosphere in which the negotiations were carried out was one of uncertainty and it was therefore not surprising that the discussions were the most difficult and protracted in the history of these negotiations." The discussions would not have been helped by a sharp deterioration in the nation's balance of payments position followed that year by a devaluation of the pound.

A year later, possibly the same minute taker reported: "the Price Review was conducted against a background of considerable difficulty" citing the country's economic problems. These, the report stated, had been intensified by the re-armament programme and the need to pay substantially increased prices for imported food and raw materials.

By now, the Government was beginning to discontinue its war time controls and to introduce a freer system of marketing. This gave Union leaders new problems in the field of marketing as witnessed in a later chapter.

One example of the new problems comes in 1953. The report records "a precipitate decision to decontrol eggs was announced without warning and interim measures have had to be hurriedly introduced to prevent disorganisation in the industry."

To deal with the Price Review, the top team of president, vice president, past president, chief executive and his lieutenants first of all held a series of internal meetings to determine their priorities.

Scottish price review team, 1958. Seated, the three office bearers, James E Rennie (centre), James Johnston and Alvan Milne; standing, D Scott Johnston, economist; HG Munro, general secretary and H D Brown, former Union president

Then followed another batch of meetings with their colleagues in the Ulster and the English Unions. Welsh farmers were not directly involved as they came under the English Union until 1978 when the breakaway Farmers Union of Wales was formed.

Having led one of those negotiating efforts, former NFUS president Sandy Inverarity recalled it involved six weeks of intensive negotiations in London, initially with the other Unions and latterly with civil servants in the Ministry of Agriculture offices in Whitehall.

"We would fly down on a Monday morning and go round to Agriculture House (the then headquarters of the English NFU) and go through the commodities one by one. The Government knew how much of each commodity it wanted and the idea was it could increase production by increasing support just as it could put pressure on to reduce production by lowering the guaranteed prices. There was an unwritten agreement the English Union would lead on arable crops and milk while we would do the same for livestock. Sometimes that led to quite strong discussions before we met the civil servants.

"Each week, the talks would conclude on a Thursday evening and we flew home on Friday. The Minister's aim was always to get an agreed Review as this made its reception in the House of Commons much easier and his officials were always pushing us for a settlement."

As the process drew to a conclusion, the Minister of the day came to the final meetings to thank the Union office bearers and their support teams. Although

Treasury officials were not directly involved, the Ministry staff was always conscious of the cost of the Review and this would be included when the Ministerial statement was made to the House of Commons.

The Ministerial comment was the first public view of the Review and it was quickly followed by the Union's assessment of the deal. "We were sworn to secrecy during the talks but we were always able to provide a quick reaction to the Review," said Mr Inverarity.

For example in 1954, the Union viewed the minimum price of milk as part of the Government's unwillingness to contemplate further expansion of production by this section of the industry.

Milk was again a problem eleven years later when Minister of Agriculture, Fred Peart wanted to reduce production and only put the guaranteed price up by one farthing (0.001 pence) per gallon.

This provoked a Union show of strength. A cavalcade of tractors was driven onto the streets as Scottish milk producers protested against this piece of parsimony and effigies of Peart were burned as they vented their anger. The following year the Review produced a rise in the guaranteed price of milk.

Initially, the Price Review was very much based on assured markets and guaranteed prices but a system of Deficiency Payments was introduced in 1954 with cereals the

Scottish Farmers were not impressed with Fred Peart's 1965 price review

first commodity to come under the new regime. These payments allowed farmers to sell at the best market prices they could obtain and then the Government paid out the difference between the price received and the price agreed for the commodity at the annual Price Review.

There were also other incentives for production with the introduction in 1966 of the Beef Cow subsidy; a move welcomed by Douglas Cargill, president of Aberdeen and Kincardine area.

Not surprisingly the open ended commitment to production soon brought political reaction and in 1960 Prime Minister, Harold Macmillan made clear his Government's unease at the situation. Four years later the Government introduced a scheme of Standard Quantities, which put a limit on production. Unsurprisingly, this was opposed fiercely by the Union. But this limit on the amount of money going into farmers' pockets did not prevent the topic of subsidies coming into the public eye with radical Labour politician, Nye Bevan MP introducing the phrase 'feather bedded farmers.'

There was still recognition of the importance of home food production in a Government Economic Report published in 1968. This pointed to the import saving role farming could play, estimating £220 million could be saved with a 22 percent increase in home production.

The Union's Livestock committee had supplied evidence to the Report making the following points: "the potential for improving output from the hills and uplands was grossly underestimated. Also in view of imports, we should exploit to the full the potential for exporting from the sheep industry." But little or no positive action was taken by the Government and a year later the Union president, Jack Blackley reported increased imports and a fall in farm production.

One of the most valued subsidies was the Marginal Agricultural Production grant, which paid out on an acreage basis and which had come into being in 1949 was, by the late 1950s, under pressure. The Hill Farming sub-committee initially - and latterly the whole Union - fought tenaciously to hold onto it but by 1963 it had gone.

It was replaced by the Winter Keep scheme but this brought division within the Union with president Major Iain Campbell having to reassure Council it was a sound scheme aimed at helping those farmers who needed financial encouragement to grow winter keep crops.

He was not helped by the regulations that excluded high lying milk producers nor the exclusion of barley as an eligible crop and the Union had to set up a special Action group to deal with the problems.

Such was the concern over the new scheme that the Union held a rally - described as the most effective in recent history by the president - in Inverness. The rally put pressure on the Government to increase the subsidy rates. This pressure brought some concessions but did not immediately add to the subsidy paid. The Government had to set up a special Appeals Committee to deal with the thousands of unhappy farmers. The majority of the appeals were upheld.

If that was a badly conceived and poorly implemented scheme, the opposite could be said of the Farm Improvement Scheme, which offered grants for a range of betterments including buildings, roads and fencing. Introduced in 1957, it was welcomed by the Union as an aid "for all farmers to improve their properties." The farmers took the Union at its word and thousands applied. In 1961, 18,731 applications were made and over £13 million paid out.

The early 1960s saw the Union in another battle, this time with British Railways over plans to cut the number of stations that handled livestock with the rail authorities proposing to close all but 48 such facilities across the country. George Dunlop, the convener of the Legal, Commercial and Cooperative committee called this a disastrous move. "It is appalling that the advice of the Union was not sought before such a decision was taken. It should not have been considered without the full views of the livestock industry."

Particularly badly hit by the cuts – later called the Beeching cuts after the head of British Railways, Lord Beeching - were the remoter areas of Scotland with Alasdair Mackenzie, the convener of the Crofters' committee saying it would add to the transport problems for the 20,000 or so crofters. This comment followed the first ever meeting of the Crofters' committee in March 1963. The Union has had many crofting members, almost from its formation, and the Crofters' committee was set up to deal with specific issues such as land tenure and crofting grants.

Miscellaneous

1954, Soft fruit
The Horticultural committee welcomed the decision of the Ministry of Food to provide for an increase in the fruit content of certain jams. An Order was made increasing the minimum fruit content of both raspberry and loganberry jam from 25 percent to 30 percent and of blackcurrant jam from 22 percent to 25 percent. Turnips were often used as fillers by jam makers.

1972, Name Change
Commenting on the Union Council considering dropping the word 'Union' from its title, noted journalist, Alex Yeaman, said: "The Council's decision saddened me a little. It was like turning our backs on the past and on the giants whose toil and sweat laid the foundations of what is well recognised here and in Europe as a powerful negotiating instrument."

And if that message did not hit home, he added: "Name changing is usually for unmarried ladies or for people on the run."

By the mid 1960s, along with talks on entering the then European Economic Community, it was obvious the UK Price Review system with Deficiency Payments would not survive.

And in 1973 following the departure of the Price Review, it received a eulogy in the Farming Leader.

"Say what you like about the British Price Review, call it a ritual dance and decry its excessive secrecy, it nevertheless emerges by comparison as a brisk and workmanlike attempt to find rational answers to problems that were often difficult and intricate. The results might not always be acceptable but generally they contained somewhere, at the core, evidence of some coherent policy of one kind or another."

And in a prescient final view, the writer observes, "If that emerges from what is going on in Brussels, it will be a miracle."

A stalwart of the Union

At his retiral in 1978, it was said of Harry Munro that he served 27 presidents of the Union, from Alan Grant in 1949 to Mike Burnett in 1978.

Harry came from farming stock in Ross-shire but war service with the RAF interrupted his studies at Edinburgh University. After training, he flew heavy aircraft involved in air sea rescue until the last 18 months of the war, which he spent in officer training, firstly at Fochabers and finally at Cranwell by which time he had attained the rank of squadron leader.

After completing his studies, he joined the Union as an assistant secretary in 1949. Following the previous incumbent, William Graham's sudden death in 1955 he assumed the top job.

Harry Munro, CBE

During his time in office as Director and General Secretary, one of the main issues facing the industry was Britain joining the then EEC.

In the run up to the UK's entry into Europe, he insisted that support for Scotland's hill and upland farmers had to be included in the package.

He was also robust in ensuring that Scottish farming should be permanently represented in Europe and a legacy of his persistence in this was seen last year with the 40th anniversary of the Bureau de l'Agriculture Britannique office (BAB) in Brussels.

Other pluses from his time as top Union official were on rating and fighting a real battle to prevent the intensive livestock sector from paying rates, and maybe more controversially steering members away from militancy as a form of protest.

His work at the Union was recognised with the award of CBE in 1964 and he became a Fellow of the Royal Agricultural Societies in 1972.

After his retiral in 1978 Harry and his wife, Ena moved home from Edinburgh to Argyll where he spent time enjoying one of his passions, sailing.

The Development of Marketing Boards

Having been introduced in the mid 1930s, only to be put to one side by Government controls during the 1940s, Marketing Boards came back onto the Union agenda in the 1950s.

Not only did they come back into focus again, their responsibilities in controlling production and promoting sales increased in the post war years, although one or two Boards did not survive long and others proposed never made it onto the statute book.

In 1955, the Government passed back to the three Scottish Milk Marketing Boards the monopoly of purchasing milk from producers and by 1961, milk quotas were proposed whereby every producer was to get a basic quantity which was to be about 85 percent of the farm's average production over the previous three years.

The Milk Boards' plan was that the producer would get a higher price for the 85 percent of 'liquid milk' – that milk sold for drinking - and a lower price for the remainder which would go for manufacturing into cheese or butter. At that time, only the liquid milk was covered by the guaranteed price.

However, a year later, a new method of providing a price guarantee for milk was introduced at the Price Review. This was the introduction of a "standard quantity" of milk for each of the Marketing Board areas. With its roots in the dairy industry, the Union was deeply involved in those discussions.

The selling of all fat stock - as prime cattle and sheep were then called - was strictly controlled, not only through the war years but well into the 1950s. It was not until 1954 that the Livestock committee of the Union discussed the formulation of a fat stock marketing scheme. Discussions also took place with the English NFU on the subject.

These discussions led to the establishment on 30 April 1955 of the Fatstock Marketing Corporation (FMC), which was created to pick up responsibility for buying cattle, sheep and pigs after the Government moved out of the market. By the end of that year, the Union reported, "the Corporation has already handled 1.5 million cattle units and has operated effectively as a price stabilising instrument."

Only a few years later in 1962, the Union was giving evidence to the Verdon-Smith committee, which had been established to look at meat marketing. The Union view was critical. It described the "present system of meat marketing as inadequate to achieve stable prices and efficient low cost marketing."

When the Verdon-Smith report came out in early 1964, it firmly rejected a proposal from the three Unions - Scottish, English and Ulster - to establish a statutory Fatstock Marketing scheme. However, it did recommend the establishment of the

Meat and Livestock Commission, which was duly set up in 1967 with the remit of improving efficiency in the industry. It also took over the grading work of the FMC.

The Fatstock Marketing Corporation was taken over by a company in 1962 with the Union holding some of the shares and it carried on operating until 1983 at which point the farming Unions in the UK had the opportunity to purchase the company provided farmers raised £10 million in share capital.

A joint statement by the Unions said: "It is now up to farmers to decide what they want and we hope every farmer will subscribe to the shares." The share issue was a flop and the various processing factories around the country were taken over by commercial concerns.

The criticism of FMC in the 1960s was not universal as the Pigs committee credited it with raising the standard of bacon pigs by penalising those animals that did not conform to its standards. That improvement might also have been due to the Pig Industry Development Authority, which was set up with the Union's blessing in 1957. That organisation's aim was one of raising quality standards and reducing costs of production. The Pigs committee reflected that this was their answer to Danish imported bacon.

Comment on livestock marketing at this time cannot pass without mentioning one of the biggest rifts between Scottish and English farmers. Following a period of poor prices for their cattle and sheep, English farmers called for a boycott of all livestock markets. Speaking on behalf of the Scottish membership of the Union, Sandy Inverarity, president stated he "did not want to widen the difference of opinion between the two but Scotland could not support the plan". Back came a telegram from Staffordshire area deploring this position and pointing out that "buyers of Scottish commodities would take note of this lack of support."

The Union's annual report of 1951 noted progress on marketing another commodity "with producers voting into force the British Wool Marketing Scheme." This was to convert to the British Wool Marketing Board and today it is the only surviving marketing board.

In its early years, the Egg Marketing Board lengthened the supply chain by requiring egg producers to sell to it. It then sold the eggs to packing stations but this unwieldy procedure was soon abandoned.

It still operated the subsidy system obtained through the Price Review based on a guaranteed price for an agreed standard quantity of eggs produced. The Egg Board's other work was as a promotional organisation with such catchy advertising slogans as "go to work on an egg" - a slogan made up by noted writer Fay Weldon in her advertising days.

In 1969, Union president, Jim McIntyre told Norman Buchan, Scottish Under Secretary of State that it was vital for Government to continue to contribute to the cost of support buying under the British Egg Marketing Scheme.

The same year saw Union Poultry committee convener, Jim Pearson in discussions with the Government on proposals to subsidise the transport of eggs by sea from Orkney to the mainland.

But by 1971, despite Union protests, the Government announced the end of the Egg Marketing Board. Recording its opposition to the decision, the Union noted this had occurred despite the Board having a sound record of technical efficiency and "in the face of producer protests."

One commodity that did not progress to forming a marketing board was grain, although the Cereals committee did give "considerable thought" to the establishment of a long-term marketing scheme in 1953.

A decade later, the issue of a Cereals Marketing Organisation was again discussed with Cereals convener, Angus Pattullo, saying that this would hold a register of forward contracts with growers eligible for a cash advance if the contract was held. The Minister of Agriculture, Fred Peart showed no enthusiasm for this proposal and no more was heard of it.

The Potato Marketing Board was reconstituted in 1955 with the new scheme having six district members in Scotland. The Union stressed the need for those members to keep in touch with its Potato committee. In 1961 it introduced a quota system in order to try and control production but with yields varying tremendously from year to year, it was a pretty blunt market control instrument.

Fruit and Vegetable problems

The press release with the most exotic heading to emerge from the thousands that have been put out in the hundred years of the Union came out in April 1969. It stated "Iron Curtain Cucumber Crisis."

The release related how imports of cucumbers from Romania were undercutting the home produced ones from the Clyde Valley. Tony Campbell, Horticulture committee convener was quoted as saying: "These so called red cucumbers have flooded onto our market without warning."

The 'Iron Curtain' was Winston Churchill's description of the division between the then Communist countries in Eastern Europe and the democratically elected countries in the West.

Apparently, the Romanians were after cash in order to fund their economy and were exporting their cucumber crop at well under the market price.

The NFUS press release noted success

Tony Campbell battling for horticulture

on the issue. "As a result of Union intervention, the standard duty on cucumbers of £1 per hundredweight (0.05 tonne) was raised by an additional £2."

Cucumbers were very much a minor crop looked after by the Union, which had in 1943 set up the Fruit and Vegetables committee. The Scottish glasshouse sector was dominated by tomato growing. In those days there were about 80 glasshouse businesses in Scotland and tomatoes accounted for three-quarters of their commercial production.

Even before the cucumber crisis, another more serious and damaging event took place in 1967 with the closing of the Suez Canal due to the Arab/Israeli war.

The Union pressed for the removal of the Government imposed two pence (one penny) per gallon surcharge on fuel oil - arising as a result of the closure of the main supply line - pointing out to Department of Agriculture officials that the surcharge could mean an increase in fuel costs of about 30 percent. No action was forthcoming and the cost of fuel continued to rise. By 1971, the Glasshouse committee members pointed out the latest increase would add about £1,400 per acre to the growers' costs in a full year.

More damaging, however, was the fact that Dutch tomato growers were receiving a subsidy on their fuel oil. This gave their tomatoes a strong advantage in the British market.

Ten years later, with the Dutch still using their own gas fields to provide cheap heat to their growers, the Union organised a petition against Dutch imports with

Miscellaneous

1949, Tenancy Tensions
The Agricultural Holdings (Scotland) Act 1949 received Royal Assent but the Union, while happy that it strengthened tenant's rights, still felt that the landlords had too much scope for resumption.

Livestock
The Livestock committee of the Union in 1958 considered there should be no importation of Charolais bulls until "such time as the potentialities of our home breeds have been fully investigated." However, two years later they had changed their position to one where controlled imports should be allowed.

1962, Potato Holidays
David Goodfellow, convener of the Labour and Machinery committee, was reporting that the Agricultural and Education Departments would no longer extend the practice of allowing children off school to pick potatoes. Producers were told to plan ahead for the future without school pupils.

Norman Lean, Glasshouse convener, handing a tray of Scottish tomatoes to the Dutch embassy in Edinburgh.

President John Cameron said the Scots could cope with tough competition but not unfair competition. "This is no lame duck, inefficient industry but now after years of unfair imports we need support at the highest political levels."

The UK Government made offers to help change oil heated glasshouses into ones heated by coal. About one quarter of the growers moved to other fuels but the economic damage by the subsidised imports had been done.

It was not until 1985 that the European Commission indicated it was about to take action against the Dutch

Cheap imports also bedevilled the Scottish soft fruit industry with complaints going back to 1959 on the importation of strawberry and raspberry pulp from Poland and Yugoslavia.

Glasshouse convener, Norman Lean, was involved in the petition against Dutch imports

For the next thirty years, successive Union Soft Fruit conveners complained about the damage this trade was doing to home production. In 1968, Tony Campbell said fruit growers in Scotland could save the country money by producing more and saving on the import bill.

In 1976, Alastair Massie, speaking at a conference in Brussels described the method of protection against imports as quite inadequate and a decade later John Whitehead was still reporting imports of fruit pulp from Poland as a significant problem.

But by 1992, Union pressure paid off and the EU granted more than £7 million to restructure the Scottish rasp industry. The money was used to buy mechanical harvesters but the industry had moved on to covered production of fresh fruit. In an ironic twist, these were picked mostly by East European labour.

The problem with rabbits

Farmers never had a problem with rabbits in the early days of the Union when a rabbit for the pot was part of the country dweller's diet. There were however tussles between landlords and tenants over the control of rabbits with this issue appearing on the Union agenda in the 1920s.

A shortage of meat in both World Wars ensured the rabbit population was not a big issue, although in the 1940s the Union had to deal with several incidents where soldiers in the Home Guard were accused of using rabbits as target practice.

Following the mild weather of 1949/50, when the rabbit population exploded, the Union was besieged by calls from members complaining of damage to their growing crops and grazing ground.

The problem was made worse with the Ministry of Food and Farming continuing to look at rabbits as a source of food. That year, despite problems with a national balance of payments, the country imported tonnes of rabbit carcases. When they were placed on the market, they depressed the price of home rabbits so that trapping became unprofitable.

The Union identified with the policy of the Scottish Landowners Federation in favouring the complete extermination of rabbits.

However, nothing was done officially until 1954 when the Pests Act put obligations on landowners to deal with rabbits and other pests on their land.

The annual report of the Union that year recorded: "Despite the efforts of individuals and local voluntary committees, an appreciable rabbit population exists in most parts of the country and it could increase rapidly."

But the report goes on "weakened strains of myxomatosis have already appeared in Angus, Aberdeenshire and Stirlingshire, in addition to several places in England".

The Legal and Commercial committee pushed for the spreading of the disease to be made legal but this highly infectious viral disease soon swept through the rabbit population without farmer help and most farmers thought they had seen the last of this pest.

The extent of rabbit damage can be gauged by the 1956 report from the Department of Agriculture which commented that "crop yields were up and part of this was due to the reduction in the rabbit population".

However, it did not take long for rabbits resistant to the virus to come on the scene and by 1961, the Union was supporting the setting up of Rabbit Clearance Societies with the Government offering a 50 percent grant towards their running costs.

These were initially very successful and the Union hosted several conferences for Rabbit Clearance Societies. In 1964, it was reported there were now 102 such organisations across the country. Following the 1964 conference, the Union lobbied the Government to ban the importation of rabbit meat and rabbit skins. That same conference also discussed the destruction of pigeons, hares and moles and that was the high point in rabbit control.

When the Government removed the subsidy in 1971, over half the Societies indicated they were winding up, leaving only 20 still in operation.

Miscellaneous

Rating

1966 saw the Union fight ferociously over plans to rate buildings used for intensive livestock production. It backed a member's court case on the issue and encouraged others to appeal any valuation

The Arrival of the Common Agricultural Policy

In any discussion on the issues that have faced farming in this country in the past century entry into the European Union (EU) would be at the top. The Common Agricultural Policy (CAP) now dominates the industry, shaping the decisions on crops to grow and livestock to keep. It largely determines the profitability or otherwise of farming and it brings with it a plethora of paperwork and regulation.

Now, after forty years within the EU and its predecessor, the European Economic Community (EEC), it is difficult to recall life without a European influence.

But it is not all one way traffic as over the same period; the Union has influenced policies and politicians on the challenges faced by farmers in this small country in the north-western corner of Europe.

It was back in 1957, that the first references to trade deals with Europe appeared on the agenda of the Union when discussion took place on the implications of joining the European Free Trade Area (EFTA).

Both the General Purposes and the Horticultural committee were "apprehensive of the implications of trade barriers being broken down" and a memo was sent to Government asking if Britain did join, then agricultural and horticultural products should not be included in the deal. Britain did join EFTA in 1960 and remained a member until 1973. The threat of cheap imported food never materialised because the other members, including Norway and Sweden, were not large-scale food producing countries.

In 1961, when talk started in political circles about the UK joining what was then called the Common Market, the Union Council laid down four demands.

The first was a system of firmly guaranteed prices maintained by government support where import control would not suffice. What the Union wanted was, in effect, the retention of the existing Deficiency Payments scheme.

Secondly, a system of annual reviews conducted on a national, as well as a Community, basis as the foundation for price determinations. This equated to the retention of the Annual Price Review system that had operated from the passing of the 1947 Agriculture Act.

Thirdly, the Union called for the maintenance of the principle of producer controlled marketing schemes - a call for the retention of the Milk, Egg and Potato Marketing Boards.

And lastly, having a system which ensured support for disadvantaged and remote areas, the Union noted the Common Market did not have any special support for those farming in less favoured areas.

However, on the opposite side of the negotiating table, the six EEC member states - France, Germany, Italy and the Benelux countries - wanted the abolition of the Deficiency Payment system the moment the Treaty of Rome was signed, together with the immediate adoption of the Community's existing commodity regulations. As far as farming was concerned that was the stand off position for the next five years with French President, Charles De Gaulle leading the opposition to the UK joining Europe. Only on his departure from politics did negotiations pick up again.

By 1970, following pressure from the Union in Scotland, Anthony Barber MP, at that time Britain's chief negotiator, was putting hill farming problems on the agenda of the first working session in the Common Market talks. He said the UK Government believed that there should be assistance for this sector of the industry in an enlarged community.

Other matters urged by the Union at that time included the need for an annual review in an enlarged Community, and an examination of the implications for milk, pig meat and eggs if the UK joined.

A year later, Union president, Morgan Milne announced at the annual meeting that the country would soon be asked to pass a basic verdict on the question of whether "our future as a nation would be as part of the European Economic Community."

By that time, it had been agreed that if Britain joined the Common Market, the mechanism of the CAP and full Community preference for agricultural products covered by EEC regulations would be introduced at the outset. British farm prices would be brought into line with Community prices over a five-year transition period starting in January 1973 during which time Deficiency Payments would be phased out.

After areas were consulted, the view from the General Purposes committee was: "The consensus of opinion within this Union is that, in the light of the assurances obtained by the Government in the pre-entry negotiations, we should be prepared to face up to the challenge of joining Europe."

Prior to the UK's entry into the Common Market, the Union invited Dr Sicco Mansholt, Commissioner for agriculture in the EEC and architect of its farming policy, to visit Scotland.

Morgan Milne led the Union into Europe

> In the past forty years there have been eleven EU Agricultural Commissioners and of that number the Union has hosted visits by nine of them. All of the visits have taken in some of this country's extreme farming challenges in the hills. The two Commissioners who did not make it to Scotland, a Latvian and an Italian, were both in office for only a short period and neither had to guide through a reform of the CAP.

After Mansholt's visit, he declared there was no difficulty in national support grants linking in with support decided on by the Community; a first for co-financing.

On the retention of marketing boards, he was more guarded saying: "The boards can stay but we will have to see whether some of their special powers can still be executed." As it turned out, when the marketing boards were dismantled, it was the UK Government that operated the wrecker's ball.

As the whole support system was changing to one where intervention buying would prop up market prices, the Union also met the chairman of the Intervention Board (IB) for agricultural produce, Sir Con O'Neill. The IB had an important role to play in issuing licences for the import and export of agricultural produce.

It was decided that the Meat and Livestock Commission (MLC) and the Home Grown Cereals Authority (HGCA) would act as the Board's agents for beef, pig meat and cereals.

Official entry into Europe came on January 1, 1973 but there was a five-year transition period for the UK. The Union, seeing the importance of a permanent base in Brussels along with the other farming Unions in the UK, set up the British Agriculture Bureau (BAB) office.

BAB has now celebrated forty years of work at the forefront of dealing with the many regulations, directives and proposals that emerge from the Agricultural and Environmental Directorates of the Commission.

Also keeping Scotland's place at the decision making tables, two years earlier in 1971, Union president Morgan Milne had assured Council that COPA – the European group of farming Unions - would be sufficiently flexible in their attitude to allow the Scottish Union to be properly represented at commodity level. Within months of that statement, Scotland had a major role in putting forward the proposals for a sheepmeat regulation. Livestock committee convener, John Cameron and Chief Executive Scott Johnston managed to get support from COPA's Sheep Group for this proposal.

Official entry into Europe did not mean the Union could sit back as there were still many big issues to be settled. In the early hours of November 21, 1973, the Council of Ministers reached agreement on the Directive covering special measures of assistance for farming in the 'poorer areas.' This was a major success for the Union and the first mark of the influence it had in making European policy.

Former chief executive, Scott Johnston believed: "Getting Europe to accept the need to support our uplands and remote areas was a huge effort but it also provided us with our finest hour."

It was not achieved without considerable effort or expense with, at one stage, the Union hiring a helicopter to take the then EU Agricultural Commissioner, Finn Olav Gundelach from Edinburgh up into the Highlands to see the terrain. For the Danish Commissioner, more used to seeing agriculture on the more gentle slopes of his homeland, farming in the hills was an eye opener.

The helicopter trip to the hills was but one part of a much bigger campaign to ensure that this important aspect of Scottish farming was included in future Common Agricultural Policies.

According to Scott Johnston: "The best thing we (the Union) ever did was to go around Europe relentlessly pushing the case for support in the less favoured areas. Initially we met with the rebuttal that this issue could be covered under the vague promise of regional policy but we persevered."

Guiding the Union through the early years of the CAP might have been Scott Johnston's biggest challenge but this trained economist had previously also been one of the guiding hands at Annual Price Reviews right from his appointment to the Union in 1957. His fluency in French was also an asset in the tussle with the French Government over its ban on lamb imports.

"Once the Government negotiators believed they had dealt with all the big issues they were not going to stick on what they saw as a minor matter. So, in the end, the CAP was adapted to meet our demands. We had played the Scottish card and had effectively, for the first time, changed the CAP in a number of important respects."

Speaking on the entry into Europe, Jim Brown, the Hill Farming convener, welcomed the progress made but emphasised that hard, intensive work lay ahead in the crucial business of designing the British schemes under the Directive.

Then in 1974, the Council of Agricultural Ministers introduced a Beef Premium Scheme prompting the Livestock committee convener, John Cameron, to state: "We are now moving to a system which, with all its imperfections, does hold out the prospect of a rising level of returns to beef producers over the critical months to come."

Although still in the transition period, Britain abandoned the Deficiency Payment system in March 1973, although not required to do so under the Treaty of Accession.

The calf, beef and hill cow subsidies were retained but support for the end price was to be provided by the EEC system of import levies related to a guide price and also through intervention buying.

For the poultry sector, transition to full integration with the EEC was accompanied by an increase in the frequency and severity of the cyclical problems experienced by British egg producers. These had afflicted the industry since 1970 when the British Government decided to dismantle the support system for eggs, which had previously stabilised the market until then.

On January 1, 1978, the five-year transition period of accession to full membership of the EEC was completed but, as was reported in the June edition of the Leader, the damage done to Scottish agriculture was not through commodity proposals but the Green Pound agreement.

Green Pound

Official UK entry into Europe did not come without its own set of problems. The biggest of those was undoubtedly the "green pound" issue.

In order to protect farmers from the daily fluctuations of the currencies of the various European countries, the European politicians had agreed fixed exchange or green rates.

Unfortunately for farmers in the UK, almost throughout its existence from 1973 until the arrival of the euro in 1999, this green rate disadvantaged them badly.

Miscellaneous

Milk to Beef

In order to reduce the amount of milk being produced, the EU introduced a "milk to beef" scheme in 1973. A sum of £15.75 was paid for every one hundred gallons of milk being produced on condition the dairy farm moved to beef production. In the first two months of the scheme, 130 Scottish farmers applied but the Union felt the scheme was primarily designed to discourage milk production rather than positively to encourage more beef to be produced. The 'milk to beef' scheme was to return later under the title of "Mulder rights" as part of an appeal on milk quotas.

Tied Houses

Showing a united front in 1973, Michael Burnett, convener of the Labour committee and Henry Crawford, secretary of the Scottish farmworkers' section of the Transport and General Workers' Union demanded the retention of tied houses for farm workers. Their joint statement came in response to a review into the pluses and minuses of this system of housing.

Politicians in this country were caught between re-valuing the rate, which would have increased the price of food, and angry farmers seeing their industry destroyed by imports made cheaper by advantageous currency rates.

On Tuesday September 10, 1974, a mass meeting with more than two thousand people took place in Edinburgh with newspapers reporting "Angry Farmers March on City" and "Food rationing could start within months, say Scotland's farmers unless the Government immediately pumps cash into the farming industry."

The march through Edinburgh to St Andrew's House came to a climax with Union president Sylvester Campbell proclaiming: "If the public want a plentiful supply of wide-ranging quality food from home sources, then they must pay for it."

Sylvester Campbell addresses members on the dangers to food supply

The following year president Fraser Evans and his counterparts in the English and Welsh Unions took part in a mass rally in the Central Halls, in London attended by more than 3,000 farmers where the issue was the damage caused to the farming industry by cheap imports.

Two years later, a trenchant article in the Farming Leader magazine on the green pound said: "It had started as a minor administrative device to avoid following the day-to-day vagaries of the money market but it has been used quite shamelessly by the British Government to subsidise the British consumer and to deny the home farmer that equality of opportunity with his continental colleagues which was supposed to inspire the whole Community concept."

In 1978, the UK Government under Jim Callaghan with a wafer thin majority in the House of Commons was defeated when it proposed a five percent devaluation of the currency. The figure was calculated by the Government as potentially putting the price of food up by one percent but also saving the UK's cattle and pig sectors from the damage from imports from Ireland and Denmark respectively. The defeat by the Conservative Opposition who proposed 7.5 percent devaluation was largely credited to the pressure by the farming lobby on the green pound.

The rift between the green pound and reality continued to widen and by 1979 the Union under president Mike Burnett called for an immediate 15 percent devaluation.

An example of the scale of damage that imports were having on Scottish agriculture during the period when the discrepancy was at that level included butter coming in from Ireland supported with the benefit of a £370 per tonne Monetary

Compensatory Allowance (MCA). Several Scottish creameries closed as a result of those imports.

> For those who believe that complexity in Europe is a new phenomenon consider the formula for working out the MCAs for cereals. The Farming Leader in 1980 explained it as multiplying the current MCA coefficient by the intervention price and then dividing that sum by the green rate of exchange.
>
> For cattle, the equation was slightly more complex requiring the following sum to be carried out to get the MCA. Multiply the current MCA coefficient found by the percentage of the intervention price, which is itself, a percentage of the guide price and then divide by the green rate.

The financial disparity also knocked the Scottish pig industry allowing Danish pork and bacon to come into this country supported by advantageous rates of exchange. This was the major factor behind the final closure of major pig meat processor, Lawsons of Dyce in 1979 having closed its curing plant in 1973.

The poultry industry was also knocked sideways with the volume of eggs imported into the UK courtesy of the green pound disparity. In 1983, Alan Taylor, Banff, the convener of the Poultry committee complained that producers were being paid 33.6 pence per dozen while costs were 44 to 48 pence per dozen with feed costs rocketing up by 11 percent or £17 per tonne. "Government must ensure that egg imports complied with regulations," he demanded.

Cattle imports both on the hoof and in carcase form from Ireland also increased at that time aided by the financial benefits from the overvalued green pounds but more of that later.

In 1988, with no "sorting out" having been achieved, Scott Johnston went on record as saying the green pound "puts our producers at a tremendous disadvantage by subsidising imports."

Echoing similar sentiments a year later, NFUS president, Ian Grant attacked the UK Government for maintaining the level of the green pound while the value of sterling fell. A call he found himself repeating a year later.

It was only in 1993, when green rates were subjected to automatic adjustments linked to market exchange rates, that matters improved for Scottish farmers.

By the end of 1995, the value of the green pound rose by 20 percent raising farm incomes by an estimated 27 percent. With this came a surge of exports led by increased tonnages of beef and lamb heading for the Continent. A trade that boomed until the disease BSE hit the headlines but that is another story.

The arrival of the euro in 1999 consigned the green pound to the history books but the damage it had inflicted on Scottish agriculture in its 25-year life is still being felt with every sector having been impacted to some degree by cheap imports.

Worst hit were pigs, poultry and glasshouse businesses but even mainline commodities saw millions and millions of pound being stripped out of farm incomes by the disparity created by the green pound. An estimate in 1990 suggested that a full devaluation of the green pound in grain would have increased cereal prices by £20 per tonne, pushed sheep values up £8 per head and raised the fat cattle price by almost £100 per animal.

Import and export problems of the 1970/80s

The knock on effects of entry into the Common Market saw the Union involved in two major battles; one against the importation of cattle and the second over the export of lamb.

The lairage at Merklands in the heart of Glasgow was built in 1907 for the importation of cattle. These mainly came from Ireland but, at various times including just after the Second World War, rangy store cattle came in from Canada to be finished on farms in the North East, the East and the Borders of Scotland.

The Irish cattle came in on boats owned by Burns & Laird who dominated the Irish Sea trade. Most of the imported cattle were quickly moved from the lairage by rail to markets in the East of the country.

At least they were moved quickly after 1936 when Anstruther Branch brought to the notice of the Central Executive that cattle landed at Merklands Wharf had to be driven three miles to Whiteinch Station, causing injury with a possible loss from pneumonia. It was pointed out that the Railway Company's line passed the Wharf and that complaints could be met if the Railway Company made a loading bank there.

With that sorted all was well until, in November 1974, Union members considered that Irish cattle were undercutting an already depressed home trade. For four days

For four days and nights, in a series of shifts, farmers from many different parts of Scotland stood at the entrance of Merkland Docks and called on those involved in the Irish cattle business to put a halt to the trade

and nights, in a series of shifts organised by a specially formed Action committee under the chairmanship of Allan MacKay of West Renfrewshire Branch, farmers from many different parts of Scotland stood at the dock entrance and called on those involved in the Irish cattle business to put a halt to the trade.

Typical of those who made the trip to Glasgow was a bus load from Aberdeenshire organised by Bert Maitland, Inverurie, vice chairman of the area. Union president, Sylvester Campbell, addressed the assembled throng.

After hearing assurances that the Union would raise the matter with the Government, the protesters departed. The Union action was successful and an import duty of £14 per head was introduced soon after.

What was not officially recorded was that the Union director general, Harry Munro had revealed to Campbell as they drove through to Glasgow together was that he had brought a crate of whisky in the boot of his car for the protestors just in case they were unruly.

Miscellaneous

Drought Problems

In the mid-1970s, two consecutive very dry summers really upset the potato market. Prior to the universal use of irrigation, yields were well below normal and growers were, in the words of the Farming Leader "well rewarded, due to the high market value of their crops." However, the report added: "the high price in conjunction with the poor quality, especially of stored potatoes, has resulted in a drop in consumption nationally to two thirds of its usual level."

Winter Warning

At the July 1979 Council meeting, speaker after speaker emphasised the plight of hill farmers after two long and expensive winters. All agreed that hill farmers needed extra assistance now, and that the promised autumn review would be no solution to their present problems. Speakers expressed their annoyance that the recently elected Government had let them down. On the election trail, Scottish Secretary of State, George Younger had said: "We cannot stand idly by, while the whole economy of the uplands is threatened." Later that year, Union president John Cameron, after praising the Government for raising the Hill Livestock Compensatory Allowances, admitted the increased subsidies would not immediately restore the economic position in the hills and uplands.

There was no such trouble. Reflecting on the restrained action by the farming pickets, Scott Johnston was told by the chief police officer who had been involved in more than a few violent protests and bloody strikes, "We do not mind you country lads coming in to picket."

If that short action helped sort out the importation of cattle, the other battle over the export of Scottish lamb took many years before it was resolved.

During the transitional period for entry into the EU, part of the deal was that lamb from New Zealand could still come into the UK in fairly large quantities. But French farmers saw the subsequent UK exports of lamb to the Continent as helping to destroy their home market.

The problem started in 1978 when the UK became a full member of the Common Market. French farmers set up very effective pickets at the import docks and managed either to close the docks or to turn back lorry loads of live lambs heading for slaughter on the Continent.

The following year, the EEC Court ruled the French action illegal. This was followed by the four UK Unions agreeing to fund and send out a consignment of lamb to test the effectiveness of the Court decision. This lorry load of lamb arrived in Cherbourg where the French duly rejected it.

Union president, John Cameron then led a "dignified and orderly" delegation to the French embassy in Edinburgh. He was accompanied by George Anderson, vice president; Patrick Gordon Duff Pennington, the convener of the Highlands and Islands sub-committee and his vice convener, Donald McNiven; Crofting committee convener, Sinclair Scott; Scott Johnston and Bob Sandilands, chief executive and deputy chief executive respectively of the Union.

The French blockade was followed up by a special meeting of the EEC Council where the French counter claimed that the UK was not allowing imports of milk from France. The tussle continued into 1980 when the Union considered taking legal action against the French for loss of sales estimated at £20 million per annum. At this point Union

John Cameron leads the Union delegation protest at the French Consulate

vice president Barclay Forrest declared: "We are approaching a major constitutional crisis in the Common Market."

Action was promised in Brussels but the problem did not go away with the French continuing to obstruct imports. In 1984, the battle re-ignited when two lorry loads of lambs were kidnapped. This direct action followed another move by the French to stop the imports by claiming the residues of sheep dip found in the fleeces were illegal in France. John Cameron was despatched to Brussels to sort out the problem but as has happened in other subsequent enforcement actions in Europe, the matter rumbled on for a number of years without any real penalties being paid.

The whole issue of exports of live animals had been on the Union agenda for some time prior to the French lamb problem. Exports of cull dairy cows had started in the mid-1950s with the Dutch buying them prior to selling them on throughout Europe. This export trade was banned in 1973, despite the Union claiming that 'free trade was free trade.' Two years later, a major effort by the Union, sustained at all levels from Branch to Headquarters paid off when, at precisely 10.13pm on January 16, the House of Commons voted for the resumption of live animal exports.

Miscellaneous

Tax Troubles
Jock Hunter, convener of the Legal and Commercial committee, was quoted as saying in 1981: "The Union threw all its resources into fighting this issue. It is no exaggeration to say that the Scottish farming industry has been saved literally millions of pounds. A major cause of concern which was bound to affect the future of the landlord/tenant system in Scotland has been removed."
His oratory was based on the Union's successful campaign to have the transfer of tenancies removed from the threat of Capital Transfer tax.

Inflation Consequences
In 1980, with inflation running at more than 20 percent - the highest level ever experienced in this country – farmers found they were losing out in the price increases. President John Cameron, speaking to the October Council, said: "We know from official figures that average ex-farm prices in August this year, were up less than 5 percent over the previous August but, in the same twelve month period, inflation increased all prices by an average of 20 percent and average farm input costs will not be too far from that figure. Little wonder that we have discovered, following our hill farming review, a drop in net incomes of around 20 percent of those farming the hills and uplands."
The following year, the Union raised its subscriptions by 20 percent.

Quotas, Set Aside, Food Promotion and the Environment

One of the main policies in the Treaty of Rome, which established the Common Market in Europe, was freedom from hunger for those within its boundaries. For those living little more than half a century after the Second World War, it is difficult to recall that tens of thousands of people starved to death in these post war years as the Continent struggled to pull itself back together again after the conflict.

Therefore the initial policies of the CAP were based on producing as much food as possible, which was good news for European farmers. Even in 1973, when the UK joined Europe, the emphasis was on growing more crops and keeping more livestock. If there was oversupply then the market was propped up with surplus production being kept in intervention stores. Soon these stores, including a number in Scotland, were bulging at the seams. The negative effect for the farming industry for this over production was that phrases such as 'wine lakes' and 'butter mountains' came into the public domain.

Something had to be done to curb the amount of produce coming off farms and soon quotas were being used to curb milk production. With cereals, land was taken out of production with a scheme called Set Aside. The politicians also decided to set a ceiling on the financial cost of the CAP moving away from the previously open-ended financial commitment.

Initially taking land out of cereal production and putting it into Set Aside was voluntary and the policy was welcomed by the Union. In 1987, President Ian

Ian Grant led early battles into Europe

Grant said: "Clearly we welcome today's statement which means that an arable Set Aside programme will be available and in operation for the next cropping year. The Union has campaigned long and hard for policies aimed at providing additional measures to help bring arable production in the EEC more in line with demand." As far as the Union was concerned the Set Aside programme was preferred to "a continuing programme of unadulterated price pressure."

Concern had been growing since the 1980s about the amount of cash going into the CAP with an article in the Farming Leader magazine putting the Union position at that time. "We should not complain and we do not complain that much public attention is focussed on the CAP. It costs around £12.5 billion per year to operate. It is right that spending at this level should be subjected to the most rigorous political scrutiny. The public, politicians and farmers themselves must continually appraise the developments of the CAP to make sure that acceptable objectives are being pursued and that money to finance the policy which incidentally accounts for only 2 percent of public sector spending in the ten member states of the EEC is being spent at least as properly, prudently and effectively as the other 98 percent."

By 1989, as further proof that the open-ended production policy of Europe had come to an end, the Union was fighting a proposed headage limit on sheep. A Council meeting that year saw speaker after speaker opposing the idea with both Livestock convener, John Ross and Union president, Ian Grant describing the Commission's proposals as lacking sense, lacking logic and discriminating savagely against Scotland. The message that Ray MacSharry, EU Agricultural Commissioner, took back to Brussels was, according to the Farming Leader: "There are hundreds of farms in Scotland's hills and uplands where one thousand ewes are needed to give a family any sort of reasonable income. These people, contrary to current Brussels mythology, are not all dukes and earls. They are just people making a modest living on some of Europe's hardest land."

MacSharry also used his visit to Scotland, courtesy of the Union, to promote his idea that the Set Aside programme should compulsorily take 15 percent of the eligible area out of production. Producers who did so would be entitled to a reimbursement of the Co-Responsibility Levy (CRL) paid on sales during the current year. The Council of the Union said it was willing to look at this proposal.

The following year, the MacSharry proposals were adopted as the new Common Agricultural Policy. However, the Union fought for and achieved major improvements to the original proposals. These were, the cereal CRL would end that year and the milk CRL in 1993. In addition, there would be no headage limits in either the Suckler Cow Premium or the Sheep Annual Premium schemes.

By the 1980s, the Union was dealing with a far wider range of issues other than those thrown up by the CAP. An increasingly well fed and affluent population was becoming more concerned about the quality and provenance of food, as well as the state of the countryside, where new UK legislation was beginning to impact on farming for the first time.

Food Quality and Promotion

When rationing was in place in post war Britain, the main concern of consumers was the actual supply and quantity of food available. However, after ration books were tucked away into history and shops became increasingly well stocked, buyers became fussier over what they were buying. Tripe, tongues and trotters, which were acceptable to a previous generation, began to be thrown into the butcher's dog food bin. Fruit and potatoes with slight blemishes were discarded as supermarkets made it clear they only wanted shiny, bright, spot free produce on their shelves.

This trend gave farmers the challenge of improving the quality and ensuring the provenance of their produce and it was one to which they had to respond.

The first positive action by the Union towards promoting food produced on Scottish farms came in September 1967 when Morgan Milne, Livestock convener, headed up a new organisation called the Scottish Quality Lamb Association (SQLA). The idea was simply to get more money back for lamb and to increase returns for everyone in marketing. The ten original members of SQLA were the Union; Thomas Borthwick & Sons from Galashiels; Buchan Meat Producers from Fraserburgh; Central Scotland Beef Producers from Cupar; City Meat Wholesalers from Hawick; the Fatstock Marketing Corporation; Scottish Meat Producers from Gorgie, Edinburgh; A Smith & Sons, Lockerbie; G D Vivers & Sons, Annan and West Cumberland Farmers from Whitehaven.

SQLA hit the ground running. In October that year, it won the Silver shield at the Smithfield centenary show and was visited by HM Queen Mother whose estate in Caithness had provided some of the prize winning carcases.

The levy of the new organisation was one penny (0.4 pence) per head with this money going towards promotion and administration.

Chief guest at the 'launching luncheon', of SQLA, was John Macintosh MP for Berwick and East Lothian, here, he is shown lamb dishes and cuts by Morgan Milne, chairman of SQLA and convenor of the Union's Livestock Committee. Also in the picture is Vivienne Fraser, feature writer for Annabel

Coincidentally, some forty years earlier in 1936, a Mr Tait speaking at a Hawick Branch meeting expressed concern about competition from imported beef. He suggested some scheme might be brought forward towards advertising their Scottish products in order to increase demand. How much would this cost, he was asked? "A penny per head" he replied. This was the first suggestion that farmers might actually promote their own produce.

A more common attitude to food promotion in the early days came in 1962 when Gordon Baxter of Baxters Foods speaking at Perth Area of the Union related a conversation with a farmer where he was told: "It is our job to produce the food not to sell it."

However, that very same year, the Union threw its weight behind a Scottish week promoting food grown in this country. Much to the amazement of Glaswegians, as part of the Union effort for this project saw a cow milked in George Square. The Scotsman newspaper farming editor, Bob Urquhart observed the performance where the cow objected to being milked. He described it as "a sair fecht but worth it in terms of publicity."

By the 1960s and early 1970s, food promotion was seen as a method of increasing sales. Following the success of SQLA, the Scottish Quality Beef and Lamb Association (SQBLA) was established in 1974 after a proposal from Moray and Nairn Area of the Union that Scotch beef also needed promotion. Part of its proposal was that the new body should not just represent producers but also include auctioneers, processors, hauliers and butchers.

In 1976, at the first SQBLA annual meeting, it was reported that 30 percent of the lamb produced in Scotland had gone to the Continent while five years previously, very little lamb had been exported to the French market. SQBLA chairman, Morgan Milne commented: "It shows what can be done with the effort of a promotional organisation and the very active wholesale trade we have in Scotland." In 1988, Milne, still as chair of SQBLA, described the pioneering role of NFU

SQLA puts on an impressive display of Scotch Lamb at Smithfield Show

Scotland in the creation of the red meat promotion body as one of the undoubted success stories of the Union.

SQBLA operated until 2008 when it was changed into Quality Meat Scotland, a Non Departmental Public Body answerable to the Scottish Cabinet Secretary for Agriculture. Despite the name change, QMS continues the work of promoting red meat; most notably under the PGI recorded Scotch Beef and Scotch Lamb labels.

The Union also played a wider role in food promotion, supporting the establishment of Food From Britain in 1982 with Douglas Cargill, Union Pig committee convener as deputy chairman under the chairmanship of Nicholas Saphir of Kent Produce.

The same year saw the setting up of the Scottish Farm and Food group with Union president, John Cameron speaking in support of the new body which had within its membership, the Scottish Milk Marketing Board, SQBLA, SAOS (representing the co-operative movement), the Institute of Auctioneers, the Council of Scottish Agricultural Colleges and the British Farm Produce Council. Backed by that combined weight of bodies, some 20 Scottish firms exhibited their wares at the SIAL international food fair in Paris.

Then in 1988, Maitland Mackie, convener of the Union's Pig committee proposed farm assurance as a scheme to "give pigs an edge in the market." That was the first mention of Farm Assurance and it came after the food industry had suffered scares such as Salmonella in eggs, BSE in beef and radiation from Chernobyl in sheep meat.

Farm Assurance then spread right through the farming industry led by the pig sector which, by 1995, could claim that all the major Scottish meat plants would only handle 100 percent Farm Assured pork and bacon.

The same year saw the establishment of Scottish Quality Farm Assured Cereals with growers having to meet stringent requirements through a voluntary code.

Chernobyl and Currie

Two unrelated explosions sent shock waves reverberating through Scottish agriculture in the late 1980s.

The first was set off by an off message politician. It lasted only a short time but caused millions of pounds of damage to the egg industry. The other, caused by mismanagement at an atomic plant thousands of miles away, brought movement restrictions that lasted more than two decades.

According to the then Union president Ian Grant, all was quiet until the ten o clock news on 3rd December 1988 when an item reported the UK Minister of Health, Edwina Currie as stating, "the majority of egg production is infected by salmonella." From that moment, he said the phone rang all night. "We were confronted with a very difficult PR exercise."

Tom Howie, Poultry convener for the Union described Mrs Currie's comments as "grossly exaggerated and alarmist for consumers" adding that some 30 million eggs were consumed daily in the UK with only an average of 36 cases of salmonella being reported. But within days of the statement, egg sales had dropped 50 percent and there was no sale for surplus birds. Mrs Currie did not even have the support of her political colleague Nicholas Fairbairn, the flamboyant Solicitor General for Scotland, reminding her that all human life started as an egg and he could see no problems with humanity.

The Union met Scottish Minister, Lord Sanderson and along with his UK counterpart, John McGregor a major advertising campaign on egg safety was launched as well as a £20 million short term buying-in programme to take surplus eggs off the market. The scare was estimated to have cost the UK poultry industry some £80 million and Mrs Currie resigned shortly afterwards.

Two years earlier on 26th April 1986, reactor four at the Chernobyl atomic power plant in the Ukraine overheated, throwing clouds of radioactive caesium into the sky. From there, prevailing winds and heavy rainfall brought the potentially dangerous chemical across Northern Europe and onto almost 10,000 sheep farms in the south west of Scotland, Cumbria and Wales.

From the records...

Prime Minister Visits

Prime Minister, Margaret Thatcher twice visited the Scottish NFU headquarters during her term of office. The first meeting was at Grosvenor Crescent and the second at the Rural Centre. At the first meeting, she talked for almost one hour to Union leaders about the problems facing agriculture and about its opportunities. In response, president, Ian Grant reminded her that agricultural net income had gone down from £200 million in 1976 to £43 million in 1985. In the same period, he added, bank borrowings had gone up £200 million to £900 million.

Union president Ian Grant described it as having to deal with a completely unknown quantity. "We had to deal with Government about how we would monitor livestock to ensure the radiation levels were acceptable for slaughter. That was an ongoing process that consumed quite a lot of Union time."

Restrictions on the movement and slaughter of sheep were first put into force on 24th June 1986 and affected 2,900 farms with more than one million sheep in the south west of Scotland. It was left to Ian Grant's successor as Union president, John Ross to negotiate the financial compensation package for the testing of radioactivity in the sheep. Travelling to London on the eve of a Bank holiday, he arrived in Whitehall as the civil servants were preparing for their long weekend. "It was one of the best and easiest negotiations that I was involved in as they were obviously wanting away." Compensation rates were set for the additional handling of the sheep required by Government testing the radiation levels.

During 1987 it became necessary to re-introduce restrictions as high levels of radioactivity were detected in some of the new season's crop of lambs who had been grazing the caesium infected grass.

Farmer and Union member, Rog Wood, Auchentaggart, Sanquhar wondered as his sheep were being tested for their radioactivity what effect the clouds of caesium had on human beings living and working in the affected areas. He was duly tested and reportedly was below the dangerous level.

The restrictions were gradually lifted but annual testing continued on some Scottish farms right until 7th July 2010 when the last controls were lifted with Bobby Carruth, Union communications director, stating it was a "blessed relief" after twenty four years of controls.

Movement restrictions brought about by Chernobyl still apply on some 300 Welsh farms a quarter of a century after the nuclear accident.

Garden Festival Glasgow 1988

Scottish farming undertook its first major public relations exercise on 28th April 1988, at the 125 acre (50 hectares) Glasgow Garden Festival site on the banks of the Clyde.

The event, which ran for five months, attracted more than a million visitors and it had as one of its central attractions, the Scottish Educational Farm. At its launch Union president, Ian Grant described the event as the biggest in Scotland since the 1938 Empire Festival.

The farm was also the launch pad for the Celebration of Food and Farming Year with George Anderson, honorary president of the Union, in charge of the Scottish element of that longer promotional exercise.

The Scottish Educational Farm's planning committee consisted of John Goodfellow, Union vice president; Bill Romanis, Union press and publicity officer; Dairmid Gunn, co-ordinator of The Scottish Educational Farm Trust which had been set up to run the project; Joe Hannah, Scottish Milk Marketing Board; David

Crowe and J Campbell Graham, both from the Union; Don Stevenson and Dr Roger Evans both from the Scottish Agricultural College and Hilary Barker whose job as farm manager was to ensure the day to day running of the 'farm.'

The 'farm' had arable and soft fruit areas, livestock paddocks and a farm wildlife conservation area complete with a pond. The main building was a 480 square metre laminated timber shed that housed dairy and beef cattle, pigs, sheep and poultry. It came complete with a herringbone milking parlour behind a huge glass viewing panel. There were clear plastic pipes showing the milk travelling through them. Lord Sanderson, the then Scottish Office Minister, spotted two young Glasgow boys watching the milking and enquired if they had learned anything. One laddie replied, "Aye. Ah ken how they get the milk into the coo, noo."

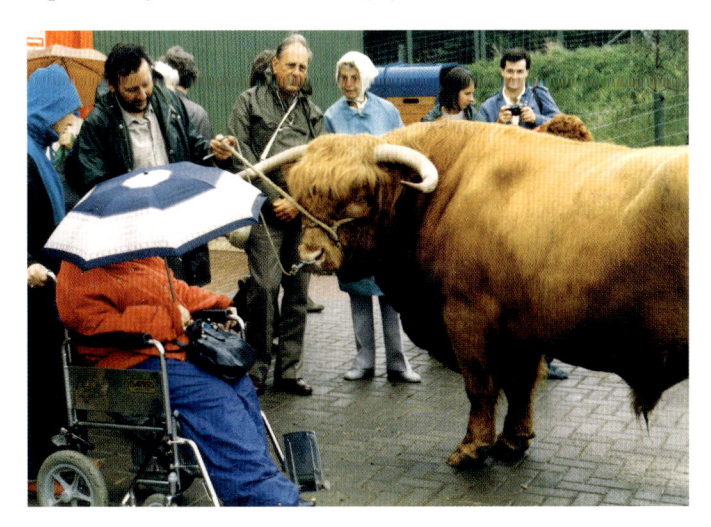

The country comes to town during the Garden Festival in 1988. Highland bull, Angus Og 2nd of Glenogle, from Pollok Country Park, meets an admirer

Union vice president, John Ross, demonstrates hand shearing to a crowd of Glasgow school children

During the five month run, there were a series of events including sheep shearing, horse shoeing and displays of farm implements and machinery, modern and old, all loaned by the farming community.

Union members along with Young Farmers and representatives from the Women's Rural Institute helped staff the 'farm.' The stewards guided and informed the hundreds of visitors streaming through and "managed" the children around the static tractors - a task in itself.

The farm re-introduced a public who had lost its links with the countryside to life down on the farm in the 1980s. Visitors came from all over Glasgow, Scotland and beyond – as far away as Japan, Hilary remembers. There were droves of school children as well as folk from farming and associated industries with the crowds particularly heavy at milking and feeding times. Children and adults were able to 'pet' and help bottle feed orphan lambs. This was a big treat for anyone enjoying hands on contact with farm animals for the very first time.

How did this 'country come to town' event go down? Comments in the visitors' book include, "the pigs are best," "smelly but super," "pure dead brilliant," "a great show" and a surprising one mentioning "good looking stewards"

The farm had its share of VIP visitors, including Prince Charles and Lady Diana passing through on the opening day. Later Prime Minister Maggie Thatcher along with her husband Denis visited as did Union president Ian Grant, who more recently in 2013 chaired the committee overseeing the building of the Hydro concert hall on the opposite side of the river Clyde. Another visitor was Union vice president John Ross who showed his skill at sheep shearing.

Hilary Barker admitted the logistics of putting it all together had been challenging but added: "Hopefully it achieved its aim of raising people's awareness of the essential contribution that Scottish farming makes to the nation."

The Forestry/Farming conflict

The Scottish Government's current ambition to dramatically increase the acreage under trees to one quarter of the total land area is not the first time the conflict between farming and forestry has been ignited in the past century.

In the late 1970s and 1980s, large areas of Scotland were planted with big tax advantages for those investing in forestry. Particularly badly hit with this rush to invest in woodland was the Flow Country in Caithness where celebrities including Sir Terry Wogan and Sir Cliff Richard became involved in tree growing.

Because much of this boggy land was not used for farming, most of the Union concerns were in other parts of Scotland where tree planting was effectively killing off hill sheep farming.

In 1979, Jock Hunter, convener of the Legal and Commercial committee, worked out that for every 20,000 hectares of trees planted there would be a loss of about 8,000 tonnes of sheep meat per annum. At that time, the long term ambition of the Forestry Commission was to increase planting in the UK to 40,000 hectares per annum giving almost three quarters of a million hectares more trees by the year 2000.

An earlier push for more tree planting came right after the ending of the Second World War. The Government charged the Forestry Commission with buying up farms on which to plant trees with the sole proviso that Department of Agriculture permission was needed before planting took place.

The Forestry Commission had originally been set up in 1919, when it was realised that Britain had few of its original forests left and vast acreages were planted in the 1920s. At that time, with farming in recession, there were few complaints over the loss of land.

But the post Second World War push for more trees saw the pace of planting rise, encouraged as it was by considerable fiscal rewards. By 1961, Union minutes show Robert Grieve from Roxburgh questioning Government policy on forestry as "the interests of farming and forestry were too divorced."

The scale of planting is revealed in the annual Department of Agriculture report for 1964. An area of 42,485 acres of land was planted in some 89 blocks of land and those figures were not exceptional for the decade.

Five years previously, the Union had expressed its concern over the loss of productive land because of increases in forestry. The Union's view was that the targeted plantings were "too ambitious."

By 1966, the conflict between farming and forestry was fully alight with an article from C Gair, Easter Moniack, Kirkhill, Inverness in the Farming Leader magazine fearing a Second Highland Clearance if forestry was allowed to continue to expand. His article referred to "sterilising the land" with trees.

> **Some thirty years before mobile phones, iPads and WiFi became common place, Mr Gair also wondered how long it would be before we would be able to "read the news on a small reading magnifier held in the pocket or handbag?"**

A month later, the Leader carried a response article from the Earl of Dalkeith, who pointed out that the country was importing £500 million worth of wood annually as the UK was only 10 percent self sufficient in timber. He also wondered how long it would be before there was a National Forestry Union with a similar remit to the National Farmers Union?

Harry Munro, Union general manager, tried to bring peace by saying that while forestry could bring additional income, it was a "very uncertain market." He added that there were social issues as well as economic ones to consider with forestry.

There was still tension two years later in 1968 when a Government report on forestry was published. It stated that "forestry should be conducted on land suitable for it but of less value than agriculture." It continued that forestry was currently being encouraged to expand in a way that made antagonism and bitterness inevitable and concluded "a dramatic change in attitude is needed."

In the decade from 1963 to 1973, the Union worked out that in that period, more than 300,000 acres were moved from farming into forestry. For comparison, the Kingdom of Fife covers 330,000 acres. This transfer of land followed approximately 150,000 acres going into woodland the previous decade. In 1972 alone, some 92,307 acres were, as the Department Annual Review notes, placed at the disposal of the Forestry Commission.

A year later in 1973, the Farming Leader observed that Lord Dundee, the president of the Scottish Landowners Federation, had put a notice on the Order Paper of the House of Lords calling attention to the need for a long-term policy on forestry. Noting that the Whips had put it down for debate on 20th June, Lord Dundee wondered if the date had been specifically chosen because it was hoped the Scottish peers who would be troublesome would be at the Highland Show and that the English peers who would be a nuisance would be at Ascot.

The conflict between farming and forestry continues to the present day.

Environment and Islay Geese

The environment did not feature much at all in the first half of the life of the Union with battles on imports, marketing, production and animal diseases dominating the agenda.

But there were exceptions with Selkirk County reporting in 1928: "Our committee has always been actively interested in measures for the better protection of the lapwing (plover), one of the best feathered friends of the farmer and a picturesque addition to the charms of rural life. It is noted with satisfaction that this bird and its eggs are now fully protected in this county all the year round. But the efforts in its favour must never be slackened 'till the selling of plovers' eggs is made illegal and full protection is given to it in the counties adjoining the shore."

More than half a century later, in the first national move by the Union, when it set up a small group in 1984 to "formulate and articulate a comprehensive philosophy on environmental issues." The group under the chairmanship of Union vice president, John Hay looked at issues arising from the 1981 Wildlife and Countryside Act, which included the establishment of Sites of Special Scientific Interest. From that time to the present day, the Union has had to deal with a massive number of environmental policies, Codes of Practice, Directives and Regulations.

One of the first environmental issues to be raised with headquarters came from Islay in 1982 when the problem of the damage caused by large numbers of Barnacle geese grazing grassland was raised by the local Branch. At that time, there were reckoned to be some 18,000 geese feeding on the 100 holdings on Islay. The letter the Union wrote at that time to all MPs pointing out the problem, referred to the goose population twenty years previously as being "just over the 6,000 mark; thus giving a threefold increase in the bird population in two decades." The letter added that the increase in numbers had occurred despite unrestricted shooting and it put the damage caused by geese to the farmers at around £60 per acre. What the letter did not mention was the solution from the RSPB who suggested appointing an official bird scarer.

The early efforts of the Union were based on increased culling but, with Barnacle geese being moved into the top priority of endangered species, this method of control did not succeed.

In the late 1980s, a report commissioned by the Union concluded: "there is a need to compensate commercial farmers adequately for the shortfall in grass and in silage yields caused by the grazing of the geese."

In a success for the Union as reported by Henry Murdoch, convener of the Legal & Commercial committee, a grant was obtained for farmers affected by the geese grazing. This was further strengthened when Union chief executive Andy Robertson took Ross Finnie, Scottish Parliament Farming Minister over to Islay in the early years of the twenty first century. A result of the site meeting saw the first land management contracts drawn up giving financial support to those affected by the damage.

The problem has continued to escalate and early in 2013, it was estimated there were now 47,000 Barnacle geese using Islay as their base and future forecasts see this rising to reach 100,000 by 2022. Backed by the Scottish Government, the Union has taken the issue to Brussels hoping to get some relaxation to the stringent regulations on culling Barnacle geese.

While Islay farmers have borne the brunt of this particular problem, those farming in a number of other locations in Scotland have also had to cope with increased numbers of geese on their land.

From the records...

1987, Scotgrow

The Union organised the Scotgrow exhibition at Ingliston with 120 exhibitors. Adam Train, convener of the Glasshouse and Nursery committee called it a "new beginning" with Scottish horticulturalists coming together for the first time at a national event

Animal Diseases

Right from the first years of the Union, one of the main issues on its agenda has been animal health. As has been noted, Diseases of Animals was one of the initial committees established by the Union in its early days as it coped with the problems brought up from the Branches and Counties.

Some of the diseases dealt with by the Union not only impacted on the animal's welfare and the farmer's economic viability but were also a threat to human health. One of the most significant of those was tuberculosis where there was a link between dairy cows carrying the disease and the human form of the disease where children suffered and often died after becoming infected.

Some of the diseases on the Union agenda at the beginning of the century such as foot and mouth are still around, others such as Aujesky's disease in pigs have been wiped from the worry list through a vigorous campaign backed by the Union; others including sheep scab were defeated but sadly have made a comeback; while more recently there are diseases such as Bluetongue and Schmallenberg that have wafted in from the Continent.

A few of the major animal diseases that have featured on the Union agenda follow but they are far from being the full range of livestock health problems dealt with in the past one hundred years.

Tuberculosis

Half a century ago, small schoolchildren may not have realised the importance of the TT mark on their one-third of a pint bottles of milk but these letters signifying Tuberculin Tested marked a major breakthrough in human health. Prior to milk having this assurance, bovine tuberculosis was one of the main reasons for TB being endemic in the human population and for the large numbers of young people who died from the disease. In the 1930s, it was estimated that 40 percent of all the dairy cows in the country had bovine TB and that there were some 50,000 cases of human TB confirmed annually; many of the latter being treated in specially built sanatoria.

The very first meeting of the Union in 1913 heard how the eradication of tuberculosis in dairy cattle could prevent young children dying from the human form of the disease. But it was not until the annual meeting in 1937 that a scheme was announced to encourage the establishment in Scotland of cattle herds officially free from tuberculosis. The scheme, with its origins in the Milk Act of 1934, provided financial assistance with a bonus of three pennies (1.2 pence) per gallon.

The Union threw its weight behind the voluntary scheme. However, the leaders of Glasgow Corporation, where there was a major problem with tuberculosis, craved Parliament to grant them the powers to enforce the compulsory pasteurisation of all milk sold in the city.

Dumbarton Branch of the Union opposed this push for compulsion on the grounds that such drastic action was likely to involve considerable hardship to the many producer retailers in the area. The branch called on the County Executive to support their view and it was unanimously decided to strenuously oppose the order by representation to the Secretary of State for Scotland and to all Scottish MPs.

The move towards compulsory pasteurisation was dropped with the Second World War on the horizon but it was quickly picked up after the conflict was over and the post war UK Government turned its attention to improving the nation's health.

In 1947, Government officials at the Department of Agriculture had discussions with the Milk committee and the Livestock committee over the early introduction of a TB eradication scheme. The plan was to start it in the most favourable areas of the country and progressively move outwards from there – a practice that is to this day used in disease eradication schemes.

One year later in 1948, the Union recorded that the Shetland Isles had gained the distinction of being the first area in Great Britain to be officially recognised as free from TB. This achievement actually came two years before the official launch of the campaign on 1st October 1950 on the mainland.

Again, thanks to inputs from both the Union and the three Scottish Milk Boards, there was a financial bonus for those going through the testing regime as well as full market value for those cattle taken out and slaughtered. Within two years of the launch in 1952, the Union was able to proclaim: "Great strides have been made in eradicating this disease (TB) from the national herd with 90 percent of the milk production now clear."

Another three years down the line and 1955 saw the South West of Scotland declared clear and 71 percent of cattle in the rest of Scotland being tested. By 1958, the North East of Scotland was the last remaining area in the country still to be declared free from the disease. This situation had arisen because the Department of Agriculture had targeted the areas with large numbers of dairy cows as a priority. There were always fewer TB problems in the beef sector.

However, with progress in eradication came problems and in 1959, with an expectation that the whole of Great Britain would be clear of TB the following year, Government announced its intention to reduce the compensation for reactors to 75 percent of their market value. The Union successfully urged the politicians to maintain the rate at 100 percent of market value.

Currently Scotland holds an official TB Free status despite there being a few cases reported every year. This is in stark comparison with the situation in England and Wales where combating TB is one of the top priorities of the Union.

Foot and Mouth Disease

Although the 2001 foot and mouth outbreak was by far the most significant disease encountered in the one hundred-year history of the Union, the virus was endemic in the UK throughout the first half of the last century. Thus, more than half the Union's sixty presidents have held office at a time when foot and mouth outbreaks were experienced in the UK. Many of the outbreaks in the early part of last century were localised with far less movement of livestock than currently exists.

Throughout the one hundred years, the main control policy has been slaughter as soon as the disease has been confirmed followed by incineration of the carcases. But as early as 1922, after a widespread outbreak with 2,691 farms taken out in the UK, the Union reported that the Ministry no longer had funds to carry out a slaughter policy; the First World War having bled the Treasury coffers' dry.

This lack of cash resulted in an isolation policy being put into operation in order to control the spread of disease. Arising from this policy, an area meeting in Berwickshire reported a suggestion by W Walker of the Rawes. He put forward the idea that where an outbreak had occurred in a field, a man with a gun should be employed to ward off hares and birds till the sunshine had time to "clear the infection from pastures." This he claimed was an extremely desirable precaution against the spread of foot and mouth disease.

It was following this outbreak that discussion first started on the option of vaccination against the disease but this was rejected and a policy of restricting beef imports from South America was promised by the then Government. Ironically, the Government Animal Health Disease centre at Pirbright, Surrey - which was to feature eighty years later in the 2007 outbreak - did go ahead at that time and produced foot and mouth vaccines which have subsequently been used in many countries outwith the UK.

In order to prevent foot and mouth coming into Scotland in 1967, Union members disinfected traffic at the Border

Only two years later, in 1924, during another outbreak in Kinross-shire farmers left their cattle grazing longer than normal because of a later summer. They then ignored movement restrictions by bringing their cattle back into their steadings for winter. The Board of Agriculture took the farmers to Court where they were defended with help from the Union. The sheriff dismissed the case and sent the farmers home with an admonition after saying there had been a "want of organisation" on the part of the Board.

The 1924 outbreak could have had benefits for the Union as Thornton Branch reckoned: "with the abatement of the foot and mouth disease, a strong effort is to be made to rope in all farmers who are still outside the Union, as it is felt every farmer and smallholder should become connected with the organisation, and those outside should get no peace until they join."

The following year after yet another outbreak - this time at Arngask in Perthshire - the local Branch pressed for a resumption of the slaughter policy on account of "the approach of the Highland Show and the hardships that would fall upon adjoining farmers in the scheduled area."

There were minor outbreaks during the Second World War but mostly these were localised and soon extinguished, the exception being an outbreak in 1942 with more than 100 cases reported throughout the UK.

The Union War committee called the outbreak "one of the most serious developments of war-time livestock production." The finger of suspicion on the cause of this outbreak pointed to the careless use of swill; a remarkable similarity to the cause of the 2001 outbreak. The War committee pressed for the introduction of more rigid safeguards in the handling and boiling of swill.

Another bigger outbreak occurred in 1952 with the slaughter of 7283 cattle, 12240 sheep and 1672 pigs in Scotland alone. The compensation for loss of stock amounted to £686,717; a figure which the Union considered "little more than adequate considering the loss of long established herds and flocks."

Outbreaks continued into the next decade throughout the UK and while most of those were down south, Scotland suffered with a major outbreak that started on a farm in Aberdeenshire in November 1960. By the time the disease had run its course in January 1961, livestock on some 104 farms in Banff, Aberdeenshire, Fife, Ayrshire, Dumfries and East Lothian had been slaughtered. Thirteen farms were affected in Orkney in the worst outbreak of the disease ever to hit the islands. The Union's Farming Leader magazine reported on this outbreak where the compensation bill for lost livestock exceeded one million pounds for the first ever time as "another instance of this dreadful scourge".

Then in 1967, following a number of years with a very low incidence of the disease, a cow with foot and mouth was identified in Oswestry, Cheshire. The problem for the authorities was that, only two days before being picked up, the cow along with 7,000 others had been to the local market. The disease soon spread from its Cheshire base to the South West of England. The whole of Scotland was declared a controlled area. To help control the disease in England, an overwhelmed Ministry of Agriculture pulled in the army; an exercise repeated in the 2001 outbreak.

On 12th December, Union Livestock convener Morgan Milne, with the intention of keeping the disease out of Scotland, asked potential festive season holidaymakers to consider whether they should come to Scotland. If they have to come, he asked that they keep off farmland and leave their dogs at home. "The spread of foot and mouth into the winter playground of Scotland's hills could play havoc with the livelihood of our hill farmers," he warned. The disease ran on for ten months in England with 2228 farms losing their stock. With just under half a million cattle, sheep and pigs slaughtered, the outbreak cost the UK Government some £26 million.

Scotland had escaped that epidemic but in sympathy with their colleagues in the South, the Union organised a support fund with Scottish milk producers being urged to donate a half penny (0.01 pence) per gallon of their milk cheque to English colleagues. On the average dairy farm this donation would be equal to £5 the Union noted. Also, to help restock stricken farms, the English Union set up a register of animals available. Scottish Area and Branch secretaries were given full details of the scheme so those farmers with stock to offer could submit details to Union headquarters.

In the aftermath of the outbreak, the UK Government set up the Northumberland committee to look into preventing any re-occurrence of the deadly disease. Submitting their views to the committee, the Union targeted a policy aimed at complete freedom from foot and mouth. But when the report came out Sylvester Campbell, convener of the Union's Livestock committee commented: "faltering courage has caused the Government to take the easy alternative of permitting the import of boned-out beef instead of accepting the basic recommendation that all supplies from endemic countries should be banned."

The Northumberland report was one of the first documents that 2001 Union president Jim Walker consulted after learning that foot and mouth disease had, once again, made an appearance in the UK.

But by that time the disease, having been identified in a pig slaughterhouse in Essex, had passed through Longtown market - the biggest sheep market in Europe. It moved quickly from there to hundreds of farms in Cumbria, North East England, Dumfries and Galloway, the Scottish Borders and further afield.

The Scottish Union role in controlling the 2001 outbreak was pivotal with Walker having daily contact with the UK Prime Minister, Tony Blair. No move to control the disease was more pivotal and more brutal than the decision to carry out a cull of livestock in

Jim Walker led the fight against 2001 outbreak

order to create a 'firebreak' in the spread of the disease. The original 'firebreak' would have meant that all stock south of the Forth/Clyde rivers were taken out but this was reined back to culling all sheep within a three kilometre radius of a known outbreak.

The whole contiguous cull for Scotland was planned over one weekend including the purchase of the burial sites. While it was "the hardest decision" he had ever made and he received a great deal of personal abuse for it, Walker believed it was the right course of action. "We in Scotland were rid of the disease in six weeks. In England they were still fighting it six months later."

The 2001 battle with foot and mouth saw tremendous efforts put in by both Union headquarters staff and local Area and Branch secretaries. Michael Coutts and his team in Castle Douglas along with Jim Milby in the Dumfries office and Ian Macpherson in Stranraer all had a difficult time dealing with members who had lost their livestock. To this day, they are reluctant to talk about the many times they were called upon during that period.

At headquarters, chief executive Ed Rainy Brown with stalwart support from policy manager, Sarah Bradburn, in addition to fighting the big issues also produced daily news sheets for members many of whom were isolated on their farms not knowing if, or when, their stock would be affected. At the local level it was often left to the Secretaries to break the bad news to farmers and to help those who had lost their stock and therefore their livelihood.

The repercussions of the 2001 outbreak which, in UK terms, cost the lives of seven million cattle, sheep and pigs and £8 billion from the public purse are still being felt with increased livestock movement controls being but one of those.

And it was the extent of those movement controls that proved the most frustrating for Jim McLaren, Union president in 2007, when the highly infectious disease escaped from Pirbright research station. All the cases that year were in the South of England but the outbreak was estimated to have cost Scottish agriculture some £32 million in lost revenue through transport restrictions.

Aujesky's Disease

As an example of the positive role the Union has played in eradicating a major disease from these shores there can be few better than the despatch of Aujesky's, that was sweeping through the national pig herd in the early 1980s.

This viral disease which causes respiratory problems, twitching of muscles, walking in circles in mild cases and abortion and mortality at the other end of the spectrum had first been identified and made a Notifiable Disease in this country in 1979.

But it was following the confirmation of a case at Heddon on the Wall - later to feature as the starting point of the 2001 foot and mouth epidemic - that the Union decided to push the Government for action.

The Government claimed they had no cash to fund a slaughter policy and placed the onus on producers to do so. The initial estimate was that it would cost £4 million to eradicate it from the UK and if producers paid 30 pence per pig this would meet the bill. A poll produced an 86 percent vote in favour of the proposal with the Ministry of Agriculture holding the kitty.

But as Pig committee convener, Fiona Dalrymple recalled, once the programme was underway the spread of the disease was far worse than initially thought. The Union also realised it was indicative of a future disaster if the slaughter programme was not carried through.

By 1983, one third of a million pigs, including thousands of breeding sows, had been slaughtered and the original cost estimate was blown 'out of the water.' Importantly, the Government was persuaded to help fund the work. By 1985, the disease was in retreat but the then Pig committee convener, Maitland Mackie had to appear before a House of Commons Select Committee to ask the Government to write off the £14 million shortfall. He was successful in this request. The last confirmed case of Aujesky's was in 1989 but it remains a Notifiable Disease in this country.

Swine Fever

Swill was also responsible for the last major outbreak of swine fever in this country. With 242 cases in 1962, the Union pushed the Government to bring in a slaughter and compensation policy. Pig committee convener, Allan Taylor said that if this was done, then "this troublesome disease" could be banished from Scotland. A compulsory slaughter policy was brought in on 11th March 1963, 20,000 pigs were slaughtered and the disease was effectively eradicated. The Pig committee considered asking for a ban on feeding swill on account of the danger of re-introducing the disease.

BSE

The biggest single bombshell to hit the UK agricultural industry in the past century came from the House of Commons. At 3.31 p.m. on 20th March 1996, UK Government Health Minister, Stephen Dorrell announced the possibility of a link between Bovine Spongiform Encephalopathy (BSE) and the human brain disease Creutzfeldt-Jakob Disease. To add fuel to an already blazing fire, he then suggested that all the cattle in the country might require to be slaughtered in order to get rid of the problem.

Within hours of the statement, tabloids were running with stories of millions of mad cows in the food chain and within days, as Prime Minister John Major recalled in his memoirs, beef consumption had plummeted by 90 percent.

For Scotland, three quarters of all farmers had some involvement in the beef sector. It was then estimated to be worth £560 million annually – 20 percent coming from exports. It was a disaster.

Within a week, the European Commission had placed a ban on all beef exports from the UK. NFU Scotland backed a legal challenge to the EU move made by their English colleagues but the decision stood.

The whole problem had emerged almost a decade earlier when BSE, or Mad Cow Disease, was first identified in cattle in November 1986. Following a gradual increase in numbers of infected cattle, it was made a Notifiable Disease in June 1988 and causal links were made to meat and bone meal in animal feed. The following month, a ban was imposed on ruminant feed being included in animal feedstuffs.

Some 182 infected animals were identified in Scotland in 1989 but this doubled the following year. Coincidentally, compensation rose from 50 percent to 100 percent in that same period.

The terrible irony for farmers was that by the time Dorrell made his statement, the number of cases of BSE had already peaked with 2203 cases in 1993. By 1995, the annual total was down to 661.

Epidemiology experts predicted the incidence of BSE would resemble a bell when placed on a graph and they were accurate in that by the year 2000, Scotland recorded only 38 cases.

Union president, John Ross accompanied by Bob Sandilands from Union headquarters told the UK Government's Agricultural committee in June 1990, the big problem for the beef industry was persuading consumers that beef was safe.

In a desperate measure aimed at assuring the public that beef was safe, the UK Environment Minister, John Gummer was photographed feeding his daughter Cordelia with a hamburger. pic Sadly this publicity stunt did not help the recovery of beef consumption.

Whenever BSE was mentioned on television, they showed the same clip of an infected cow tottering about on concrete as if she was trying to walk across an ice rink.

Looking back, Union president, John Ross said BSE was a major issue throughout his term of office. "One difficulty we faced saw one part of the industry blaming another. We had to move away from that and get the various players together in order to restore confidence."

His term of office as president was completed some ten days before Dorrell's announcement leaving newly elected president, Sandy Mole, and chief executive Tom Brady to deal with the UK Government's self inflicted explosion of concern over Mad Cow disease. Within two hours of the announcement, the new top team was on its way to London to speak to UK Ministers and to present a series of drastic measures to help alleviate the problem.

John Ross had to confront BSE

Chief among the proposals was the introduction of an Over Thirty Month Scheme (OTMS) which, before it was ended more than a decade later, would see more than two million cattle all more than two and a half years old being consigned to the incinerator.

The rationale for this scheme was that the virus was much more likely to be harboured in older cattle. The cost to the public purse of the OTMS scheme was just under £2,000 million; some £800 million of this was recoverable from the EU.

In June 1996, the Union backed by the UK Government did persuade an emergency meeting of the Agricultural Council of the EU to introduce a series of measures including raising the support for the Beef Special Premium scheme and the Suckler Cow Premium scheme.

Sandy Mole fought many battles to gain compensation arising from the BSE debacle

This was followed in October with more support going into the Calf Processing Scheme and additional monies to help promote beef sales.

It would be another two years before any Scotch beef made its way across the Channel. Even then, the conditions attached to the consignment were onerous with details of every animal having to be supplied.

An inquiry by the Auditor General into the costs of BSE found that some £1,576 million was spent by Government in the financial year 1996/97 trying to get stability back into the beef industry. The final Government figure for the BSE crisis was put at just under £4 billion following a later inquiry.

But no figure was ever put on the loss of income and damage to the farming industry by the Mad Cow scare. However it is certain that without the Union fighting the cause, the damage would have been far greater.

To this day, the BSE legacy lingers on in a number of ways notably in slaughterhouse legislation. Some export markets are only now opening their doors more than a decade after the BSE storm blew through the UK meat industry.

Grass Sickness

Grass sickness in horses is still a major problem for the equine world but when horses provided the power down on the farm, the disease was a major economic worry for the farming industry. In 1939, Perth Area asked headquarters to "Impress upon the Highland and Agricultural Society and the Department of Agriculture the urgent need for offering a money prize, of substantial amount, to the first person who might succeed in finding the cause of, and remedy for, the disease."

Sheep Scab

Having been eradicated in 1952, sheep scab re-entered the country in 1973 and since then has spread rapidly to a point where it is now endemic. In order to bring this parasitic problem under control, the Union helped support the 2010 Sheep Scab Order, which obliges owners of infected sheep to report when their animals have the ailment.

Ninety years previously, the legislation produced by Government required the dipping of sheep twice per annum in specified areas with this work being carried out under the eagle eye of the local policeman. Initially the dipping requirement did not meet universal approval in the Union with Dumfries County recording in their annual report in 1927 its congratulations to headquarters that members had been freed from the "burdensome and oppressive conditions of the threatened Order."

Despite this opposition, sheep scab was eliminated from the country by the early 1950s.

Fowl Pest

This disease was first identified in this country in 1926 with an outbreak in Newcastle – hence its alternative name of Newcastle disease. The Union first became involved in dealing with it in 1949 when more than 200 cases were reported; most of those were located in the Western Isles and Orkney.

Outbreaks continued to occur throughout the 1950s; in 1956 the outbreak was triggered by an importation of parakeets from Belgium. The Union's role at his time included battling for appropriate compensation for infected flocks.

The biggest outbreak occurred in 1970 with the Farming Leader magazine reporting the destruction of 64,000 birds and a compensation bill of £90,000 being paid out.

Brucellosis or Contagious Abortion

Having been an intermittent problem for years - both on farms and for those working with cattle - the UK Government decided in the mid 1960s to introduce a scheme to eradicate brucellosis. When the plan was presented to the Livestock committee in 1965, members wholeheartedly supported it and two years later it moved into law.

Progress in removing brucellosis from the country was slower than some newspapers would have liked but in 1973, Union Milk committee convener, Henry Christie claimed great progress stating,

"More than 72 percent of registered milk producers in Scotland have accredited brucellosis herds and another 15 percent are undergoing tests to achieve accreditation. This makes a total of 87 percent accredited or on their way to it. Compare that with a figure of only seven percent fully accredited four years ago and you will see how seriously the industry is tackling this problem."

Other diseases
Animal health is now a devolved issue and is the responsibility of the Scottish Government. The Union is currently working with the Government on the planned eradication of Bovine Viral Diarrhoea. This is one of the most costly diseases for cattle producers and it is estimated that 40 percent of Scottish herds have BVD infected cattle.

Recent changes in weather have also seen some diseases such as Johne's and liver fluke spreading across Scotland and becoming economically significant. The Union has been working with research institutes and the State Veterinary Service in developing control and management policies for these problems.

As the Union steps out into its second century, the emphasis on animal disease has moved towards prevention. As such, it has worked with the Scottish Government on policies aimed at preventing disease coming into the country.

Summary
These are just a few of the major animal disease problems the Union has faced in its one hundred-year history. To list all the battles fought on the livestock health front would be tedious. The ones that have been mentioned give an idea both of the importance of animal health to the farming industry and the time the Union has devoted to dealing with the consequences of infection.

There is no end to this battle. There is far more travelling across continents both by humans and animals than there has ever been. These movements are being carried out in very short time spans. Both these realities help spread disease. Changing weather patterns are also bringing new exotic challenges on the animal health front and the threat of animal disease to the Scottish farming economy might be greater now than it was a century ago.

Weather

The fortunes and failures of farming in Scotland are inevitably closely linked with the weather; a truth faced many times in the past one hundred years by the leaders of the Union.

One of the earliest examples of the close relationship between the profitability of farming and the fickle nature of the weather came in the annual report Dumfries area sent to the Union's National Executive in 1928. The words used in that minute could easily be transposed to similar weather disasters throughout the past century. "While all farmers would fain forget the year 1927, history will record for the information of future generations that it was one of the worst and most trying seasons within memory. All will hope for better and brighter things in 1928. Even those - and there were many - who felt they were competent to advise farmers as to how their business should be conducted had to confess that the conditions in 1927 were hopeless and they were really sorry for the plight the industry was in. In no other trade or profession are such difficulties met with or is more patience required."

Because there can be no influencing the weather, the role of the Union down the ages has been to mitigate the disaster, either financially or through practical help.

The one certainty, as stated by former Union president, Ian Grant is that no one should become president of NFU Scotland unless he, or she, is prepared for the consequences of a totally unexpected spell of weather.

Nigel Miller, president who experienced the 2012 so-called summer and sodden autumn followed by the very late cold snap in the spring of 2013 confirmed those sentiments.

All those who have guided the Union in the past decade with its more unpredictable weather may feel that they have been buffeted more than most by gales, cold snaps and flooding as the weather appears to have become more extreme. But right down the last century, Grant's words have proven true with extremely cold winters, equally extreme wet summers, as well as flooding and drought, all coming along to make farming in Scotland exasperating or interesting, depending from which angle it is observed.

While the Union cannot influence the weather it has, particularly in the past forty years, built up a reputation for gaining Government aid following climate disasters. Prior to that, Union action was mainly in coordinating relief supplies into weather affected areas.

In weather terms, Ian Grant had two difficult years. Both of them seriously affected silage making and equally importantly harvest crops. The 1986 disaster seriously hit the west coast and the central belt of Scotland while two years later it was the North East that suffered.

On the first occasion, following continuous rain over much of the country with rainfall from July to September being 200% to 350% above normal and sunshine less than 75% of normal, a presidential visit was made to a number of dairy farms.

First stop was to Killoch, Galston the home of one of the most famous Ayrshire herds where the Woodhead family had to take their 170 milking cows indoors from 24 July. Winter feed was therefore being consumed at a rapid rate and, as Ian Grant and Union general secretary, Scott Johnston heard, was "only likely to last until January".

The presidential team then visited Jim Allison at Dunsyre where they were informed that no hay had been made because of the wet weather. The team went on to Ayrshire where they visited Bill Shankland's farm at Crosslar, Cumnock where again no hay had been made and he had 100 dairy cows to be fed.

A 'fodder scheme' with Area and Branch secretaries compiling registers of those who had and those who wanted hay and straw was launched.

The worst affected areas were Wigton, Dumfries and Stewartry, Ayr, Lanark and Inverclyde. By the end of the winter, the Farming Leader reported "thousands of tonnes of fodder have been transported to the affected areas."

On the political front, an urgent message was sent to the Prime Minister and to the Secretary of State for Scotland as well as a series of crises meetings being held with the agricultural colleges, Members of Parliament and importantly - bearing in mind the negative effect on cash flows - the clearing banks.

The result of

President Ian Grant and Union leaders survey the storm damage in 1985

this Union pressure came in December with a £7.73 million rescue package worth £14 per cow and 35 pence per ewe in affected areas. But when the farm income figures were published for that year, it was revealed that incomes had dropped 75% on the previous year and the bad weather had knocked £100 million off balance sheets.

Two years later, severe weather at harvest time hit the North East and Ian Grant was commanded to attend a meeting at Auchnagatt. The room was packed with angry farmers with one overexcited agricultural reporter recording "they are hinging from the rafters."

Grant's recall of that meeting was just before the meeting started the manageress came in and said, "Excuse me I have to make an announcement, would the owner of car registration number xxx kindly remove it because it is causing a blockage."

From the registration number Grant recognised that it was a new car and so when he stood up he said he was comforted to know there was at least one farmer who was able to buy a new car despite the difficulties. This 'off the cuff' remark broke the ice but it did not remove the seriousness of the situation.

Again part of the solution was a call to Government. Initially, Scottish Minister for Agriculture, Lord Sanderson responded by saying the industry had a tendency to talk itself down into spiral of downward confidence. Then he pointed to various Government schemes that were supporting the industry.

However, once he grasped the gravity of the situation, he used his influence as a former director of the Clydesdale Bank which had a major share of bank borrowings in North East to see how they could ease the situation. According to Grant, Sanderson also instructed the Department to look at non-bank borrowing to get a truer picture of the situation and this move was followed with a private meeting between the Union and John McGregor, UK Minister of Agriculture.

Ironically, those two episodes of dealing with wet weather had followed a dry year where according to the Farming Leader, the Union president had highlighted the high level of culling of dairy stock because of the drought conditions.

The first involvement of the Union with the consequences of bad weather however go back into the First World War, when in 1916 appeals were made to the Board of Agriculture for increases in the prices being paid because of the severe weather in gathering in the crops.

Bad weather also interrupted production in the Second World War with the annual report of the Union recording that in 1941/42 the severe winters of the past two years had considerably aggravated the hill sheep farmers' plight.

"It was obvious that the subsidy of 2/6d (12 pence) per breeding ewe previously awarded would have to be increased substantially before this form of assistance would have any appreciable effect on the situation."

Other extremes of weather over the past century

Snow

In extreme weather terms, the post war years were marked with the big late snow fall in the spring of 1947. With extensive damage and major feeding problems facing livestock farmers, the Union encouraged the setting up of local efforts to share fodder. But with major losses in certain districts being so great the Union also decided to join with the NFU of England and Wales in an appeal for, and the administration of, an Agricultural Disaster Fund.

The Union was also involved in 1963 dealing with what the Farming Leader described as "the worst winter since 1947." The Air Force and Naval helicopters carried out 70 sorties taking 150 tons of feed to hill farms isolated by the heavy snow. The Government is recorded as waiving the £18,000 cost of the military help.

In another snow hit year in 1982, president John Cameron announced the Department of Agriculture were ready to support Operation Snowdrop whereby fodder supplies could be dropped to snow bound hill farmers.

It was again a later unexpected fall of snow that caught livestock farmers in the west of Scotland in 2013 causing severe losses of livestock. Here again, pressure from the Union helped the Scottish Government produce an aid package for those affected.

During the late snow falls of 1947, the Union encouraged the setting up of local efforts to share fodder

Wind

Union Headquarters estimated that the widespread damage caused by the hurricane which hit central Scotland on the night of 14-15th January 1968 had resulted in losses pic to farmers and growers of at least £3 million.

Worst hit were the owners of glasshouses who suffered badly. In some cases growers had their houses demolished. Following a meeting with Norman Buchan, Under-Secretary of State, Alastair Campbell, convener of the Legal and Commercial committee welcomed a statement by the Department of Agriculture that they would give priority under the various grant schemes to gale damage repair work.

A decade later, hill farmers in the Highland and Grampian regions were badly hit by the blizzards which ravaged much of the country at the end of January. Some farmers sustained losses in sheep numbers of between 25% and 50%. The Hill Farming sub-committee convener, Patrick Gordon-Duff Pennington and

Many agricultural sectors are hit hard when there are high winds, however, the owners of glasshouses can literally see their livelihood disappear in front of their eyes

Union office bearers met the Department of Agricultural officials and, as a result, a compensation fund was set up.

Following a lot of effort by head office and group secretaries, Scottish Power paid out £25,000 to Union members in recognition of the difficulties caused to dairy farmers in the region after the Boxing Day storm in 1998.

In the twenty first century, storms in 2000 and 2005 caused widespread damage with the NFU Mutual paying out £20 million in the first of those and £13 million five years later.

Drought

In 1977, John Cameron who was standing in for Potato committee convener, Barclay Forrest, announced that the Department had decided to lift the ban on exporting ware potatoes. This ban was first announced on 26th September 1975. It was one of the measures taken to mitigate the shortfall in production from the short crop that year following the hot dry summer weather. The ban continued in force during the 1976 season when similar conditions prevailed.

Flooding

In 1967, parts of Ross-shire suffered serious flooding. Subsequently a Union team met Sir Mathew Campbell of the Department of Agriculture and relayed the gravity of the problems facing individual farmers. They asked for re-instatement of the flood banks and "adequate" compensation.

Three years later on 30th August 1970, the Union president, Sandy Inverarity visited farms in Moray to make an on-the-spot investigation of the damage caused by flooding. The Union urged immediate emergency action by the Department to prevent any recurrence of the damage and to investigate what measures could be taken to mitigate any future occurrence.

A decade later, following flooding in a number of areas, the Union again pressurised the Government into providing grants of up to 70% for flood banks and arterial drainage.

Branches, areas and regions

Those who helped set up the original operating framework of the Union with Branches and Counties reporting to central headquarters would be pleased to see that the same basic communication routes are still operating today; albeit with emails and texts as opposed to letters.

The framework that has worked over the past one hundred years does so because it encourages local members to become involved and contribute to the messages going to headquarters while at the same time providing a method of getting the latest news from the top.

The devolved nature of the framework means that every area's priorities and specialities can thus be dealt with. The agendas for a meeting in the arable areas of the country are totally different from those in the livestock producing hills and islands and yet the information from all areas can be gathered together to help produce the policies the Union works to.

For some members of the Union, the local meetings provide the information they need to help their businesses. For others who want to move into public life, local meetings provide a stepping stone to higher office. The ambitious know that all sixty Union presidents started by attending their local Branch meetings.

While the framework remains largely the same, the Union commenced a rationalisation of its Branches and Areas on its fiftieth birthday in 1963. It then abandoned Areas in 2000 combining them into larger Regions apart from Ayr, which has retained its identity throughout.

Underneath is a short summary of the changes along with views and stories from local representatives.

ARGYLL AND THE ISLANDS

The first minutes of the Argyll Area show that on 10th January 1919, a meeting was held in Campbeltown to decide whether a Branch should be formed for Kintyre. Some 27 farmers signed up and were followed by another 17 over the next two meetings.

Argyll (Cowal) was called Dunoon Branch for some time. The split between North and Mid Argyll did not occur until 1939 but separate Branches operated at Lorn and Lochgilphead from 1919. The North Argyll Branches in Mull and Tiree are of more recent vintage. In South Argyll the pattern of Carradale and Campbeltown obtains from 1918 while the county's fifth Area of Islay, Jura & Colonsay was formed in 1928.

Arran, now a one Branch Area, started with Branches at Southend and Kilbride and in 1919 formed a third Branch at Shiskine. In 1934 Arran, by this time reduced to one Branch, came under the wing of Ayr Area but it recovered its Area status in 1947.

Re-organisation in 1967 saw the Branches of Campbeltown and Carradale go, with them forming a new Branch called Kintyre.

By 1990 Mid Argyll was a one Branch Area with 78 members while North Argyll had four Branches and 272 members. Bute and Cowal were both one Branch Areas with 74 and 60 members respectively. Islay, Jura, Colonsay & Gigha formed a one Branch Area with 75 members, as did Kintyre with 158 members.

When Argyll and the Islands was declared a Region in 2000, ten Branches remained – Lorn, Sunart, Kintyre, Arran, Mid Argyll, Tiree, Mull, Bute, Cowal and Islay, Jura & Colonsay.

Comment on the region comes from Ronnie McLauchlan from Tulliemet: "The North Argyll Area which is central and meets in Oban must be one of the largest and most difficult to service in Scotland. It is made up of the Branches of Lorn, Mull, Sunart plus Tiree and Coll. Lorn is a district from Arduaine, which is about half way between Oban and Lochgilphead in the south, to Ballachulish and Glencoe in the north, and east to Dalmally, plus the islands of Kerrera and Lismore. Mull Branch covers the whole island and Sunart Branch includes Moidart, Morven and Ardnamurchan; all to the west of Loch Linnhe. This was covered for most of my time by secretaries Archie MacIntyre, who lived at Minard on Loch Fyne and latterly Trevor Bainey who lived at Dalmally. Both were excellent and served Argyll well as secretaries and NFU Mutual agents."

Commenting on some of the characters from the area, he relates: "Lorn Branch had an early claim to fame in producing a Union president, Major Ian Campbell of Arduaine. Other characters were his brother, Sandy Campbell, Achnacreebeag and Alex Fleming, Fannans who was a pioneer of hill land regeneration which included fencing his farm into blocks and rotating his cows and ewes, New Zealand style, to make grazing more efficient. Brigadier R Fellows, Claddich was a well-known Highland Cattle breeder, while Neil MacDonald, Barguillein was another land improver and Ann McLarty, Glenshellach, our only lady member, was a well-known producer of in-calf beef bred heifers. On the dairy side were Malcolm McGregor, Strontoiller and Hugh MacInnes, Ariogan both from Oban and Andrew MacIntyre, South Ledaig, North Connel.

"Sunart Branch produced Betty MacDonald, Rhemore who was convener of the Crofting committee during the 1970s and 1980s. She was a very formidable lady! Captain Tony Bailey, Inversanda, who was retired from the Royal Navy, was on the Legal & Commercial committee and was a great champion of the Phelps Trap.

At that time the gin trap was still used to trap foxes but was about to be banned. The Phelps was more humane as it didn't have the brutal snap of the gin but sadly it never did get accepted. Jim MacLaren, Craig Farm, Dalmally was a Council member and a deep thinking man who always gave sound advice. He was followed to Council by his son Donald who now runs Craig with his brother Alistair.

"From Mull we had Lachie MacLean, who farmed Knock, Gruline. He was a well-known and respected member as well as being a Gaelic speaker, local historian and father of Lachlan, the present convener of the Less Favoured Area committee. Davie Galbraith, Kentallen was another of the Mull representatives to Area. He always seemed to have some complaint about the service they got from MacBraynes ferries. George MacRae from Bunessan in the Ross of Mull was a man who had led a colourful life. He had been a shepherd in the Falklands for a number of years and was also a very good ice skater, so good that he performed in the international ice show at Olympia.

"Our Area meetings were always held in the Kings Arms Hotel, Oban, opposite the bay, so if things got a bit dull which was not very often, one could watch the ferry boats arrive and leave. The meetings started at 11am with a short break for lunch and usually ended by 3pm. This allowed members to catch their ferries back to Mull and Sunart but usually not before the Kings Arms Bar was cleaned out of gin and tonic.

"I served on Council myself as a member of the Livestock, Hill Farming and Tenant Farmers committees over a number of years and got to know many fellow farmers from all over Scotland with friendships still to this day. Although we discussed some very serious stuff, the camaraderie and social side of these meetings were always most enjoyable."

AYR

Ayrshire farmers were not slow off the mark following the inaugural meeting of the Scottish Farmers Union in October 1913. Little more than two months after that event, at the George Hotel, Kilmarnock on 14th January 1914, Craigie & District Branch was formed with eminent stock farmer, James Kilpatrick, Craigie Mains as president. But Kilmarnock Branch beat them to it, as it was formed in late 1913 and had as its president the Union's own first president, William Donald, Fardalehill. In the following year, John Young, Skerrington Mains is recorded as secretary. His son, Willie Young, later Sir William, became Union president in 1948 and later he was chairman of the Scottish Milk Marketing Board. Another member of that early Kilmarnock committee was James Caldwell whose son served on Union Council in the 1960s.

Also in January 1914, the formation of the Dundonald Branch is recorded with Mauchline Branch being formed in March. Beith Branch also got under way and on its first committee was John Allan, who was still an active member when the

Union celebrated its 50th anniversary. In 1917, Dalry Branch was formed and the first minutes show a list of some 80 farmers who had 'signified their intention of joining the Dalry Branch.'

By 1918, Ayr Area had 18 Branches but in 1930 Stewarton, Craigie and Fenwick Branches disappeared. Millport on Cumbrae Branch was formed in 1934; Prestwick and Monkton Branch became enveloped in Ayr Branch in 1939 and Ochiltree Branch was formed in 1947. Long serving Area secretary, Stewart Gilmour had 17 Branches and 1950 members to service in 1949.

Reorganisation in 1967 dissolved Sorn, Ochiltree and Cumnock creating Cumnock & District Branch. Mauchline and Tarbolton gave rise to Failford. Kilmarnock and Dundonald made way for Kilmarnock and Kilmarnock West & Dundonald. By 1990, the Area had 1253 members but was down to 10 Branches.

When the end of century reorganisation occurred, Ayr was the only former Area to become a self-contained Region. The Branches were, however, cut down to five, those remaining being Failford, Kilmarnock, Cumnock & District, South Ayrshire and North Ayrshire & Cumbrae.

Providing his memories of Ayrshire, long serving Area secretary, Howard Jefferson recalls: "It was always an Area that provided great characters who carried the views of the Area to Headquarters. The Area was the second largest in the organisation (after Aberdeen and Kincardine) and quite often punched above its weight. At one point, Ayr had four members on the General Purposes committee of the Union. They were Henry Murdoch, Legal & Commercial convener, Ian Kerr, Milk convener, Adam Train, Union treasurer after serving as convener of the Glasshouse and Nursery committee and John Scott MSP was Hill Farming

Ayr members feared no one, and sent contingents to London to emphasise their point

convener. Other Ayrshire men to make their mark included Matt Simpson, Hill Farming, Robert Lamont, Legal & Commercial, David Purdie, also Legal & Commercial.

"The list also includes two great orators, Willie Campbell former Milk convener who is still active in milk politics and John Duncan who, having missed out in a vice presidential election, went on to be chairman of Scottish Milk; thus following other Ayrshire greats Bill Weir and Sir William Young.

"Two names who would have gone further in the Union if their lives had not been cut short were Ian Kerr, who sadly passed away at an early age and, who many believed, would have gone on to greater things; also Ian McMillan, a local vet and man of wisdom and humour.

"Ayr always had close links with its MPs. It sent contingents to London to emphasise a point. On one occasion, the Area 'persuaded' Headquarters to join a convoy of Ayrshire members on a trip to London when Hill Farming was having a difficult time.

"Ayr Area held monthly meetings with the agricultural press attending every one. This, at times, was frowned upon by Headquarters but proved to local members that we did have independent minds and would represent their views at all times. Quite often there would be four or five journalists at the meetings where they found out grass roots feeling on issues and when the press were requested to 'lift their pens' - they always did so. It was a strong relationship and Area and Headquarters were the better for it.

"Ayr Area members were often of an independent mind and feared no one. On one visit, president John Ross remarked that he needed the United Nations Secretary General, Boutros Boutros-Gahli to accompany him to Ayr Area."

DUMFRIES AND GALLOWAY

The early records of the Union in Wigtownshire are very full with a minute in November 1918 recording that the rather loosely knit local organisation was consolidated into a County Area. Other early shifts saw Machars and Rhins split into separate Areas in January 1919 and three years later in July 1922 the three district Branches in the Machars combined to form a one Branch Area.

In the Stewartry of Kirkcudbright Area, the first records are also dated 1918. By this time the national Union had 95 Branches in 25 Counties. The single Branch Area of Stewartry was formed in 1917 with 38 members, and their minutes of 28th January, 1918 show membership had reached 116 but also recording they thought this to be too low a percentage of the farmers in the area. The Area met nine times in 1917 and discussed among other items, the sales of calving queys, cheese colouring and grading prices for sheep and cattle.

The third Galloway county of Dumfries was also established in 1917 and by the end of the year had Branches in Thornhill, Annan, Dumfries and Lockerbie. The first available minute of the Area in May 1917 advises that, "members should have their rail fares refunded following meetings in Dumfries." The first minutes of the Annan Branch on 30th March, 1917 reveal a noteworthy continuity in the post of secretary. The first occupier of that post, J Roddick being succeeded by his son, W Roddick who was still secretary at the 50th anniversary of the Union in 1963.

In both Dumfries and Kirkcudbright, these early structures were maintained until the two Areas of Dumfries and Stewartry were amalgamated in 1974. By 1990, the combined Area still had seven Branches with 868 members and the single Branch Area of Wigtown had 479 members. With the formation of a Region in 2000 the remaining four Branches are at Langholm, Stewartry, Dumfries-shire and Wigtown.

The Miskelly award is presented every year in memory of the late John Miskelly, a a much-respected Union staff member from Aberdeenshire. The first winner of that award was Michael Coutts and the Union Chief Executive, Andy Robertson said at the award ceremony, "Mickey Coutts is a local stalwart who has borne great pressure throughout recent years, especially helping those struck or severely restricted by foot and mouth in 2001. His composure in difficult times has helped many members get through BSE, Salmonella in eggs and Stranraer port protests, all the while ensuring the Union's activities reflected members' interests and that its reputation was maintained. It is extraordinary that Mickey has time in his day to play his bagpipes far and wide, be an extraordinary dog handler, a skilled shot and angler as well as a member of the local constabulary."

Here, Mickey Coutts recalls: "The Union Area secretary at the outbreak of World War Two was local solicitor Patrick Gifford, a Spitfire Pilot who was credited with shooting down the first enemy plane over Scotland. He later went

Former Dumfries and Stewartry Area Presidents pictured with Secretary John Coutts MBE, at his retiral presentation in 1988. Pictured from back: CW Campbell, AR Campbell, TC Gillespie, I Wilson, D Hendry, PT Gordon Duff Pennington Front: Wilf Shaw, John Coutts and A V Hamilton

missing over France and neither his body nor his plane was ever recovered. There is a memorial to him in Castle Douglas. Meanwhile a partner in his family law firm of Patrick Gifford & Co. continued with the role of part time Branch secretary.

"My father worked for the law firm concerned and administered the NFU affairs until the NFU secretary role went full time in 1969. At this point, my father joined the Union, which was linked to the Mutual agency. He also assumed the role of Dumfries & Stewartry Area secretary on amalgamation in 1974. I joined him as assistant secretary in 1985 and took over as Branch secretary on his retiral in 1990."

Further west, past President John Ross provided these memories: "In the early 1960s Wigtown had two Branches, Rhins and Machars and resolutions from each had to be approved by Wigtown Area committee before they could be forwarded to HQ. After much deliberation, an outbreak of peace was achieved and Wigtown became a one-branch area.

"The number of delegates eligible to attend NFUS Council was dependent on membership numbers at year end. While there was some flexibility allowed, there always seemed to be a significant number of members in arrears. I recall travelling around Wigtownshire with the late Jimmy Cunningham, at that time NFU secretary, attempting to gather up subs. We were usually well received but didn't always get a cheque.

"Issues regarding milk were always high on the agenda in Wigtownshire. We had two high profile figures from that industry on our Area committee; Bob Lammie, chairman of the Scottish Milk Marketing Board, and Henry Christie, convener of the Union's Milk committee. With two men of that calibre, discussion could get heated, but at least dairy farmers in Wigtownshire had access to two men who could really influence policy.

"Two stalwarts of Wigtown Area Executive were Dougie McCrone and Moray Rhind. Both were heavy smokers and as the night wore on, business was often conducted in a haze of smoke. Eligibility for the then Sheep Annual Premium was dependent on retaining ewes for a qualifying period which happened to coincide with the holy period of Ramadan. Mutton was in big demand in that period but if you sold your fat ewes in the retention period, you forfeited your subsidy. Dougie therefore proposed, supported by Moray, that we seek Headquarters' support to change the timing of that festival - the motion was not carried.

"The Chernobyl fallout hit South West Scotland hard and much of the area was closed to the sale of finished lambs. Gradually, however, on a parish by parish basis areas were released from restriction. At that time, farming had a spot on Radio Scotland every evening and I recall the presenter, Charlie Allan 'phoning me in a panic regarding the pronunciation of some of these places - particularly Torthorwald and Kirkgunzean!

"The threatened closure of Stranraer Creamery demonstrated the importance of a strong Union presence locally and the value of NFU staff to co-ordinate action. The creamery buy-out team of Ian Evans, chairman, Robin Christie, vice-chairman,

George Lammie, David Caldwell, Alan Marshall, and Wigtown area secretary Ian MacPherson all did sterling work to resolve this issue. The creamery has survived to this day to the benefit of dairy farmers and the wider community. Even in difficult times, there has always been humour, friendship and a sense of purpose. And, as the NFUS moves into its second century I trust that will continue."

EAST CENTRAL

This Region comprises the old counties of Angus, Fife, Kinross-shire and Perthshire.

Fife was a single Area with three Branches in 1918 before moving to two Areas in 1923. The original three Branches of Dunfermline and Thornton (comprising West Fife Area) and Cupar (East Fife Area) had been joined by Anstruther Branch in 1922. At Anstruther Branch's first annual meeting some fifty farmers attended and Henry Watson, Drumrack was appointed president. His son followed him in 1960 and his grandson, the present Henry, made it three in a row in that office. A number of the founder members were still giving active service at the 50th anniversary of the Union including Frank Roger, Kenly Green - who was for many years convener of the Sugar Beet committee - J W Clement, the first Branch secretary who was a member of the Scottish Milk Marketing Board. H H Eadie, whose son was president in 1958, was still a regular at this time.

Cupar Branch went into voluntary liquidation at the end of 1930 but was resuscitated in 1933.

In Angus, the Branch position remained static for a long period with five Branches – Arbroath, Brechin, Dundee, Forfar and Montrose. Tay Valley was formed in 1944 to bring the total up to six.

In Perthshire in 1918, there were two separate Areas. The original four Branches in East Perth were augmented by Aberfeldy Branch in 1933. On the other hand, a reduction in Branch numbers took place in West Perth. In 1928, the Thornhill Branch, which at times was called Forth Valley, joined with Port of Menteith Branch, and in 1944 Balquhidder and Killin Branches joined forces.

By 1990, Angus had five Branches with 419 members, Fife & Kinross had four Branches with 530 members and Perth had five Branches with 806 members.

A decade later, all three of these Areas were grouped into one Region, the remaining seven Branches being at Highland Perthshire, Crieff, Angus, Perth, North East Fife, Blairgowrie, and West Fife.

Ewan Pate provided these memories from Angus: "For all its eighty years existence the Angus Executive held sway nationally on a number of topics but none more so than on potatoes. In 2013, potatoes seem almost non-political but that was far from the case in earlier decades. The Angus Executive was right at the forefront of the battle to establish a Potato Marketing Board (PMB) in the 1930s. Dundee branch member Tom Pate served on the Executive for 60 unbroken years from its foundation in 1922. He always recalled the battle for the PMB as one of the most difficult ever fought.

"The end of the PMB 60 years later was no less controversial with the Angus NFUS holding faithful to the end. Few who attended will forget the arguments presented in favour of retention by PMB Scottish Special member and former NFUS vice-president John Hay of Carlungie, Carnoustie. Not everyone appreciated so much potato talk however. One evening, dairy farmer Gordon Law from Glentyrie, at Redford exploded after sitting through a PMB report, an NFUS potato committee report and a Scottish Seed Potato Development Council report. 'It's nothing but tatties at these meetings!' he complained. 'How about something else for a change!' His cry, although well warranted, went largely unheeded. Old habits die hard.

"In fact there were plenty of other reports. Will Lindsay from Lednathie in Glen Prosen held the fort for the livestock producers for years. John Goodfellow, an NFUS vice-president and so nearly a president as his father David had been, kept his Angus colleagues up to speed with the legal niceties. Tacit relocation has never before or since been made to sound as interesting.

"The Angus Executive was peopled by many remarkable characters but few have been as enduring as Area secretary, Duncan McKenzie. Still hale and hearty in his mid- nineties and still a Forfar resident, he ruled the roost with consummate ease and bonhomie from 1950 until 1983.

'I was only the third Area secretary to be appointed in Scotland,' he recalled. 'Before then the secretaries in Angus were lawyers or other professionals. It was a great opportunity for me. After the war I had returned to banking but was torn between that and a career in the motor trade. Fortunately Alex Archie, the executive chairman at the time, persuaded me to instead join NFUS and I never regretted it. Over the years I had a ball and consider myself a very lucky man.'

"Duncan was also to be a very busy man. Angus had five Branches – Arbroath, Brechin, Dundee, Forfar and Montrose as well as a specialist soft fruit Branch known as Tay Valley. Each had a full programme of winter meetings as well as an annual dinner and in some cases a dinner-dance. In all he acted as secretary to 116 chairmen over 33 years. He achieved all this and his ever-prospering NFU Mutual work single handed. He has, however, always acknowledged that he had a secret weapon – his well-trained and ever efficient 'girls in the office.' One of them, Edna Gerrard, now Dalgarno, went on to become a popular Branch and then joint Area secretary. Duncan McKenzie's ever sharp memory recalls that there were 1000 members in Angus at the start of his tenure but only 736 by the end of it."

John Arbuckle provided these memories from Fife & Kinross: "Fife & Kinross has been well represented at Union Headquarters for most of the last 100 years. The stellar names of Andrew Arbuckle and John Cameron of course both served as president. Peter Stewart and Robert Howat both served four year stints as vice-president with John Picken doing a two year stint while Alan Bowie has served in the same post up until the present.

"Andrew Logan and John Whitehead both had a spell as convener of the Soft

Fruit and Field Vegetables committee. Andrew Peddie was convener of the Pigs committee following his father as a convener, Andrew senior being convener of the Sugar Beet committee. This committee also had Frank Roger, Charlie Samson and Andrew Arbuckle as its conveners before the factory closed in 1971. Andrew's nephew John was convener of the Potato committee and Russell Brown is now chairman of the Specialist Crops committee which includes potatoes.

"Other conveners were Harry Melville, Poultry Committee, Douglas Alexander, Labour and Machinery, Tom Thomson, Milk Producer Retailer, George Lawrie, Environment and Land Use and Iain McCrone, Fish Farming.

"Jimmy Wilson, Allan Bowie and James Adams have all served as chairmen of the East Central Region with Jimmy going on to serve as National Treasurer; a task now undertaken by George Lawrie.

"Other Fife stalwarts and regular attendees at the annual dinners include Dr Jean Balfour, Alan Barr and Bob Mitchell."

From Perth, John McLaren who served as Potato committee convener has identified a few of the prominent Union members: "David Sinclair MBE, Abernyte was a professional footballer in his younger days. He was also the youngest NFU member to become president of Perth Branch, was a member of the Area Executive and an active and enthusiastic member of the Branch all his life. He was chairman of the Potato Marketing Board and his favourite saying was, 'the only thing better for you than a tattie is two tatties.' An enthusiastic livestock showman, he was also a member of the Royal Smithfield Club and for a spell as chairman.

"Stewart Miller, Rosefield was a well-known potato grower and active member of Perth Branch. He became a Councillor on Perth County Council and gained a reputation for strongly supporting rural and agricultural matters in Council.

"Andrew McLaren, Wester Keillor was an active member of the Branch committee with strong views on the future of agriculture regarding political manoeuvring following World War Two and was a prominent producer of first class prime cattle.

"David Yellowlees, Muirhall Farm was a dairy farmer, member of Perth Branch and served on the Scottish Milk Marketing Board.

"Bob Simpson, Duchlage Farm, Crieff was also a dairy farmer, president of Crieff Branch and member of Perth Area Executive. He served on the Milk committee, the Scottish Milk Marketing Board and fought tirelessly to prevent the SMMB being abandoned, constantly warning of what the outcome would be.

"John Mathieson MBE, Inchmagrannachan, another dairy farmer was president of Atholl Branch and Perth Area. He was convener of the Legal & Commercial committee, a director of NFU Mutual Insurance Society and became Provost of Perth and Kinross while serving on their Council.

"Duncan McDiarmid MBE, Castle Menzies Farm was president of Aberfeldy Branch and Perth Area. He was convener of the Organisation committee and also served on the Livestock and Hill Farming committees. He became a member of the Scottish Land Court when Lord Birsay (Harald Leslie) was chairman."

FORTH AND CLYDE

The inaugural meeting of East Kilbride Branch was held on 14th January, 1914 and in the brief record it was agreed that committee members should "go out in pairs to canvass for new members." Another Branch to start up in 1914 was Hamilton, Dalziel & Dalserf, which is one of the longest titles in the Union history.

On 26th July 1917, Biggar Branch was formed and thus projected the Union in Lanarkshire to its most easterly extremity. In the next month's minutes the Branch's influence was extended over the county boundary to become the Biggar and Peebles Branch. Meetings were to be held alternately in the two centres.

In Dunbartonshire, it was decided to form a Branch for the Western district of the county and a Branch for the Kilpatrick district and "to recommend the idea to the favourable consideration of the farmers in the parishes of Kirkintilloch and Cumbernauld." Two points of interest from the early Dumbarton minutes reveal that show societies were in many instances instrumental in introducing the Union to various parts of the country and those early Branch committees took the enrolment of new members as an integral part of their duties.

The early records of the Mearns and Paisley Branches no longer exist but press reports of the day show that these were probably the first two Branches formed. The Eaglesham Branch first minutes of May 1914 show that the decision to form a Branch was taken in the appropriately named Union Hall.

Clackmannan or Alloa Branch was formed in 1918 and remained unaltered for many years but in Dumbarton, Kilsyth Branch in 1919 joined the original three of Dumbarton, Kilpatrick and Kirkintilloch. Cumbernauld Branch completed the picture in 1937.

Lanarkshire had eight branches by 1918 but by 1925 Wishaw, which had been formed in 1918, amalgamated with Shotts, which was established in 1919. Old Monkland Branch, which was formed in 1919, moved in 1925 to join New Monkland & Airdrie Branch. 1926 was a particularly bad year and both Lesmahagow and Biggar went into abeyance that year but both were resuscitated soon after. Biggar, which had previously been in another area in 1922, was the ninth of the Branches to join the Area. The Lanark picture was completed when Clyde Valley came into being in 1942.

In Renfrewshire the Union, quite naturally, was in good numbers branch-wise by 1918. The Cathcart and Eastwood Branch vanished in 1938 due to inroads of building eating up farmland but, that apart, the Area pattern remained that way up to the 50th Anniversary of the Union.

By 1963, Stirling Area was down to three Branches whereas in 1918 there were ten. Accounting for this reduction, Kilsyth Branch joined Dumbarton Area in 1919. Then Gargunnock, Kippen and Bannockburn merged with neighbouring Branches. Falkirk and Denny joined forces in 1928 and the Buchlyvie and Drymen Branches lost their identity in 1933 and 1952 respectively.

The 1967 re-organisation saw several changes. The Dumbarton Branches of Kilsyth and Kirkintilloch merged to form Kelvin and the new Branch of Central Scotland Horticultural was formed. The Renfrew Branches of Abbey, Newton Mearns and Neilston became East Renfrewshire and the Stirling Branches of Falkirk & Denny and Cumbernauld became Falkirk.

By 1990, Forth Valley had seven Branches with 624 members, Inverclyde had eight Branches with 500 members and Lanark had nine Branches with 807 members.

When the three Areas were made into the Forth and Clyde Region, ten Branches remained - Forth Valley, Clydesdale, Avon & Nethan, Lanarkshire North, Kelvin, Balquhidder & Killin, Balfron & Menteith, Renfrewshire, Dumbarton & Kilpatrick and Falkirk.

Long serving secretary of the Union, Charlie Shanks - from 1972 until 2003 - is a native of Lanark and his first job was with the auctioneers Lawrie & Symington. After 13 years there, he joined the Union as Area secretary in Renfrewshire and Forth Valley.

Another long serving stalwart is Mary Fisher. Mary has been with the Union since 1985. She was secretary of the Inverclyde Area and is now branch secretary of the amalgamated Dumbarton & Kilpatrick Branch.

Willie Wilson, Kelvin Branch supplied the following: "I joined the NFUS late 1967 as the new tenant of holding No 4 Barraston after meeting the then branch secretary Bill Burgess, who had been visiting a neighbouring farmer where I was working part- time. Two years later, David Ralston, another neighbour asked me to go along to a Branch meeting where there would be an attendance of between 25 and 30 members . The total membership in those days was about 150. I remember one meeting when Cumbernauld new town was about to compulsorily buy a lot of farms around Cumbernauld and there was standing room only, as members came for information and support. During a beef crisis late70s/early 80s members were rallied to picket Merkland docks in Glasgow to stop Irish beef flooding in as beef prices were very low. Our Area secretary at that time was Arthur Taylor who co-ordinated the pickets. A few years on, I was elected to go to Area and would travel with Tom Gray, Alistair Lang, Jack Brewster, David Ralston, Willie Chalmers and Donald McGregor - all working farmers able to put their point across and I enjoyed serving alongside them.

"After the sudden death of Arthur Taylor, Reg McLatchie was appointed secretary and during his tenure Dumbarton and Renfrew Areas were amalgamated to form Inverclyde. Meetings were then alternated between Renfrew and Dumbarton where larger turnouts brought more discussion and livelier debates. Some meetings could go on long into the evening, milk and livestock taking centre stage.

"New names and faces to me were Basil Baird, Willie Baird, Alex Ritchie, Robert Carruth, Robert Steven and many more not forgetting Louis Anderson, who was never short of a good debate, especially on tenancy subjects. Reg McLatchie tragically died a few years later.

"In 1985, Mary Fisher became our new Area secretary - a young woman In a man's world full of enthusiasm and energy - to support Branch and Area members. Inverclyde ended on the 29th January, 1999 but Kelvin Branch survives although with many fewer members as farms disappear through amalgamations and the former Department of Agriculture Holdings are no longer the source of many pig and poultry businesses. Kelvin Branch is still supported by a very high percentage of working farmers."

HIGHLAND

This Region now encompasses the former counties of Caithness, Sutherland, Ross & Cromarty and Inverness-shire.

Most of the early activity is documented in 1918 when Caithness, Wick and Thurso Branches were formed. They were joined by Latheron in 1936.

Also in 1918 there were Branches at Inverness, Glen Urquhart, Lochaber, Grantown and Carr Bridge although Glen Urquhart became 'lost' in Inverness & District in 1921, while the two Branches on Speyside went into abeyance after a short life. Grantown and Carr Bridge were served as a Branch from Moray Area which was formed in 1927. Lochaber Branch received no mention from 1931 onwards but reappears as an Area in 1935. In 1967 a new Branch was formed at Loch Duich in the Inverness Area.

The County of Ross & Cromarty operated as an Area in 1918 with two Branches at Tain and Wester Ross & Black Isle. In 1922 the two Branches split company to form two separate Areas and in 1924 Wester Ross & Black Isle broke up with both becoming independent Areas.

Sutherland Area began with one Branch at Golspie and it was not until 1945 that new Branches were formed at Rogart, Lairg and Bonar Bridge.

By 1990 Caithness Area had 436 members in four Branches, Sutherland had 163 members in two Branches, Easter Ross was a one Branch Area with 104 members, Black Isle & Mid Ross likewise with 228 members and Inverness had four Branches with 253 members.

Under the new set-up in 2000, Highland Region was left with 11 Branches at - Nairn, Loch Duich, Inverness & District, Strathspey & Badenoch, Skye, Sutherland, BlackIsle & Mid Ross, Lochaber, Easter Ross, Moray and Caithness.

One of the characters of the Highland Region is George Murray (93) from Sutherland who was given the Miskelly Award at the 2013 annual dinner. In making the presentation, president Nigel Miller said: "His drive and enthusiasm for farming is undiminished and, in our Centenary year, George and his family encapsulate so much of what is good about Scottish farming and the families at its heart."

George Murray remembers: "I became a Council member when Alan Grant was president in 1949/1950 and was surprised to be put on the Pigs committee and the Organisation committee. Later I was more at home when I served on the Hill

Farming committee.

When the annual meeting was held in Aviemore in the early 1980s I had to welcome the delegates. I approached it with much trepidation but was given great moral support from John Cameron, president at the time.

I really enjoyed my time with the Union but I do not want to say any more in case I am accused of blowing my own trumpet."

Jamie Grant, Black Isle remembers: "Murdo Holm, secretary of the Black Isle Branch was a whale of a character, huge fun and outrageous on occasion. He established a singing group called 'Murdo Holm and the Clod-hoppers' to provide entertainment at NFU Dinners.

"I was 23 years old when I attended my first NFU meeting in Avoch. An old stager asked me, 'So what are you going to do on the farm?' I answered, 'Grow potatoes.' He replied, 'Great crop, potatoes, you've always got something to worry about."

Comment from Highland Region would not be complete without reference to Morris Pottinger and Iain Thomson.

Morris from Caithness was on Council in the 1980s. He was not one to toe the party line but always spoke up eloquently for his Area and his commodity.

Iain was convener of the Highlands and Islands committee also in the 1980s. The story has been told that as one of the more cerebral members of Council he used to spend Council meetings reading the Economist or the Ecologist rather than his agenda.

LOTHIANS AND BORDERS

In the Border county of Roxburgh, four Branches were formed in 1918 and one in Newcastleton in 1945. An early committee member of Kelso Branch was William Graham, Marchcleugh, who became president of the Union then its General Secretary. Jedburgh Branch was established shortly after Kelso and St Boswells Branch followed after that. St Boswells had, on its first committee, R H Allan, Smailholm Mains who was still active in union circles at the 50th anniversary. Hawick Branch completed the Roxburgh quartet.

Union activity was also introduced into Berwickshire in 1918. The first minutes of the old "Middle District" are dated 18th February, 1918 and the first vice-president was J P Ross Taylor (later Sir Joshua), who became president of the Union in 1928. Berwickshire began with three Branches – East, Mid and West – in one Area but in 1924 West went off on its own to form a one-Branch Area and East and Mid did likewise in 1936.

The Union in Peebles started off with Branches at West Linton and Peebles. At that time they were part of the special area which included Biggar Branch. With Biggar leaving to join Lanark in 1922, Union activity in Peebles slumped somewhat. In fact it is not even recorded from 1924 to 1932. At that time the one-Branch

Area came back into being. Incidentally, though small in number, the Peebles Area contrived to provide two Union presidents – W D Jackson in 1943/44 and G Hedley in 1952/53.

Selkirk Area was formed in 1918 at a meeting held on 9th January with one of the committee appointed that day being Walter Inglis, Netherbarns. His son was the Area president at the 50th anniversary.

In 1967 the two Branches of Hawick and Jedburgh combined to form the appropriately named Branch of Hawick & Jedburgh.

In 1917 the Union became active in the Lothians and the minutes of the East Lothian or Haddington Branch of 16th February report that 50 farmers had offered to join. At that time one Area Executive was contemplated to cover the whole of the Lothians and after the formation of a Branch in Midlothian the third county came in when, at a meeting in Linlithgow on 29th July 1918, West Lothian Branch was formed. Twenty four members enrolled and in the minutes of 19th August, the Branch took the somewhat momentous decision to allow "reporters" to attend meetings. By 1922, Mid and West Lothian Areas had combined. The original Branch complement was four, but in 1923 Linlithgow and Bathgate Branches linked up and in 1950 West Calder lost its identity with members mostly transferring to Bathgate Branch.

In 1990, Borders had eight Branches with 896 members while Lothian had four Branches with 675 members.

Following reorganisation in 2000, the new Region of Lothians and Borders had ten Branches at Peebles, Midlothian, East Lothian, Hawick & Jedburgh, Newcastleton, St Boswells & Kelso, Selkirk, West Berwick, Bathgate & West Calder and Mid & East Berwick.

Bob Noble, former secretary in the Borders remembers his area. "The Scottish Borders, the 'debatable land' between England and Scotland, was the domain of reivers and rustlers. They were tough and independent men. Borderers are still tough; especially those in farming! So it was a challenge for NFUS to organise the eight Borders NFU Branches into one entity – The Borders Area.

"At an early Area meeting, when the arable farmers complained about their sector, the hill men stood and dabbed their eyes! The Area president, Tom Elliot, slammed the table and said this attitude would not be tolerated. The Area would represent all farmers and the Union was about farmers working together! From then on Borders farmers worked as one under the banner – 'Borders'.

"Communication was paramount whether they were Council members' reports to Area or reports at Branch level. The Area had a magazine called 'The Border Leader', which published articles written by Area and Branch officials as well as Council members. Maximum use was also made of the local press to keep members informed. Politically, regular communication was maintained with MPs and other organisations concerned with land use.

"Gradually the area grew in stature and influence. Surveys and questionnaires were sent to ordinary NFU members on important issues such as the McSharry proposals. Meetings were organised to initiate group and co-operative ventures and care was taken to ensure accurate commodity information was gathered and communicated to headquarters.

"The Area also organised a range of social and sporting events with the Area dinner featuring a list of ex Prime Ministers becoming legendary.

"Not much happened in Borders farming that did not pass through the Area's agendas but perhaps the greatest legacy of the Area is the talent it produced. Building on the inspiration of Willie Swan and Jim Stobo, the Area introduced and trained a succession of talented individuals who have gone on to serve the Union and the Industry at high level; people such as former president Sandy Mole and current president Nigel Miller.

"The Borders Area was of its time. It was fun. Developments in computers and communication techniques mean things are done differently now. But not everything changes and NFUS is still about farmers working together."

Former Lothians Area Secretary, John Biggar recalls: "I was appointed in the late 70's as secretary of Bathgate & West Calder Branch. Being a Borderer this was somewhat of a culture shock but in my early days with the then Branch president, John McIndoe and past president, Robert Young in the wings, transition to Branch and Area was made easy. West Lothian at that time was predominately a dairy area. As Secretary of the Lothian & Borders Milk Producers, there were over 130 producers within the Area. How many are there now? Not many but possibly just as much milk being produced. Milk quotas were a huge topic with many angry meetings where Henry Christie, the Milk convener at the time, faced tough questioning.

"The Branch was the first step in the debating procedure with often very strong views being passed up to Area. It was amazing when the views were received from the differing types of farming areas of Mid and East Lothian that a different light was shed and common ground was reached to pass on to head office; this did not always meet with approval at the next Branch meeting! This, coupled with reports back from Commodity committees, was seen as a very important link to and from head office and one arguably not as well informed now under the Regional set-up.

"A few years later I was appointed Area secretary with John Kinnaird as Area president. This was near the time of the demise of the Area which saddened many of us with so much common ground found with inter Branch discussion. To mark the passing of the Area, all the committee members were invited by Lord and Lady Rosebery to a wake at Dalmeny House where a cake in the shape of a coffin was produced.

"Over the years, and latterly working with all three branches, many characters, definitely brightened up these meeting with controversial comments to spark off heated debates. Robert Young, Stuart Wardlaw, Jimmy Barrie, Andrew Bathgate and Willie Kerr were all masters at this. The steadying influence of the likes of John

Kinnaird, Roy Black, Andrew Lorrain Smith, Duncan Barr and Jamie Smart was definitely required. It was also very pleasing during my time to see sons following in fathers' footsteps as Branch presidents, George and Kelvin Pate (or "Mr Paté" as George was introduced at an East Lothian dinner when he was president by the Manager of the Marine Hotel), George and John Sinclair, and John and Ian Orr.

"During my 30 years with the Union every problem seemed to be greater than the one before but all were very ably dealt with by NFU Scotland: even highly publicised items such as Milk Quotas, BSE, Foot & Mouth and the Common Agriculture Policy."

ORKNEY AND SHETLAND

In 1918, Orkney had four Branches covering East and West Mainland, Shapinsay and Rousay. These numbers were increased in 1933 when, on 19th October, Holm Branch was formed with both the Union's president Dr G G Middleton and its vice-president J A D Pirie addressing the meeting. Later that year, on 1st December a Branch was formed in the adjoining parish of Deerness with an initial membership of 44. The first secretary at Deerness, Robert Hepburn, continued in office until 1963 when, along with the other part-time secretaries in the mainland of Orkney, he made way for a full-time appointment. On 21st April 1934, Birsay Branch was formed with an initial membership of 90 and with G R T Scarth, Twatt Farm, Birsay as first secretary; he became president of Orkney Area in 1963. By that time there were nine Branches. Shapinsay continued but there was no Branch for Rousay. The outer island Branch of Sanday was formed in 1920 but Branches formed for Stronsay and Walls and Hoy in the same year went by the board in 1923 and 1940 respectively. In 1967 there were new Branches at Westray, Stronsay, Rousay, Egilsay & Wyre, South Ronaldsay & Birsay, Eday and Walls & Hoy.

The Union only came into Shetland in 1942 to form a one-Branch Area.

In 1990, Orkney had 953 members being serviced by twelve Branches. Shetland had the one-Branch Area with 223 members.

When Orkney and Shetland became a Region in 2000, ten Branches remained at – Orkney, Shetland, Shapinsay, Rousay & Egilsay & Wyre, Eday, Westray, Stronsay, Walls & Hoy, North Ronaldsay and Sanday.

George Burgher, long serving secretary in Orkney provided these memories. "The NFUS is first recorded in Orkney in 1918 and by 1922 there were two branches on mainland Orkney and four on the outer islands. J R Hourston, a farmer, held the post of secretary of the County Executive committee of the Union for no less than 33 years, during which time most parishes formed their own Branches and Secretaries were instructed by members to hold meetings during full moons!

"In 1963, when George Scarth was president, John Riddell was employed as the first full-time Area secretary and also secretary of the ten mainland Branches, which were reduced to five in 1965. Total membership soon increased from 778 to

1172 and, when George Stevenson was elected Area president in 1967, he instigated much more activity with delegates attending meetings in Edinburgh and national office bearers visiting Orkney on a regular basis.

"John Riddell resigned to start farming in 1969 and he was succeeded by George Burgher, who serviced the four mainland Branches and supported the eight island Branch secretaries. Peak membership of 1,365 was reached in 1973 when Orkney had the third largest Union membership in Scotland. The present secretary, Kenny Slater, was appointed in 1998.

"All office bearers gave freely of their time and talents on behalf of members. It is impossible to name them all but a few of the important events are mentioned below. All produce for the Union's annual dinner in 1974 was, for the first time, sourced from one Area - Orkney. President, Jackie Tait helped to find and arrange transport for all the food. More recently, in 2006 food for the dinner was provided by Orkney & Shetland Region.

"In 1979-80 Orkney suffered disastrous wet weather conditions and president, Billy Wood spearheaded the successful campaign for assistance for Orcadian farmers from the Government, Highlands & Islands Development Board, the local authority and the EEC.

"Bob Bain was Area president in 1981-2 and went on to become vice-chairman of the Union's Legal and Commercial committee. He played an active part in securing the Agricultural Development Programme for Orkney.

"During Ronnie Cursiter's presidency in 1984, Orkney Area purchased and renovated the building at 60 Junction Road to form one of the best NFU Offices in Scotland, with ownership still vested in Orkney Branch.

"Also in 1984 grants were withdrawn for slatted buildings to be used for breeding cows and Area representations led to a visit from members of the Farm Animal Welfare Council who observed cows on slats at seven farms in Orkney. This resulted in these grants being restored for beef cows where suitable bedded areas were provided for calves and calving.

"In late 1997 Donald Dewar, Secretary of State for Scotland was due to visit Orkney and feelings amongst farmers and the wider community were running so high that a spontaneous demonstration and blockade of central Kirkwall took place with 500 protesters and several tractors. This was very unusual for Orkney and scenes such as these had not been seen since the huge rally held in 1930 when 5,000 people demonstrated in Kirkwall against cheap imports. President Marcus Wood and past president Terry Coghill met Mr Dewar and stressed the serious position of farmers in Orkney.

"Travel to meetings in the south has become easier and less time consuming. Whereas, in the 1960s, a meeting would mean an absence of three days, now meetings in Edinburgh can be achieved in one day. All Area and Branch presidents and other office bearers have given much of their time working for members and none more so than Stewart Wood, who was Branch chairman from 2005-6 and who in 2007 and 2008 went on to become national vice president – the highest office in the Union ever attained by an Orcadian."

From Shetland, Davie Nicolson provided the following memories but omits to mention his own role as convener of the Crofting committee for many years. Retired chief executive Scott Johnston relates Davie was responsible for raising the membership in Shetland from 100 to 400 during the 1970s. He also remembers Davie taking him round Yell show and being introduced as 'our man from the south'.

"Robert Winton was heavily involved with the Union in the early days and was the first islander to sit on the national Crofting committee. Over the years there have been a number of prominent agricultural representatives who have taken the chair in Shetland. Many have sat on national committees such as Crofting, Highlands & Islands and Livestock. Issues dealt with locally would include the Braer oil spill in January 1993, dialogue with the local authority on local agricultural schemes, Suckler Cow Premium, livestock shipping, health schemes pertaining to Shetland and ensuring Shetland was well represented with Government and EU schemes.

"Over the last 50 years there have been five secretaries, working from their homes in the early years on a part-time basis but latterly from an office in Lerwick. In that time there has been a marked change in the number of people involved with active full-time crofting. Many have alternative employment with the croft now a part-time activity. As a result of this and current family and financial pressures, it is difficult for crofters to find the time to sit on committees such as NFUS. This, along with land transfer, has had a negative effect on membership. NFUS still has a very important part to play in Shetland agriculture; even more than ever with the current CAP reform."

NORTH EAST

The North East of Scotland has been one of the Union strongholds and it has also been a remarkable area as regards individual acts of lengthy service to the Union.

By the end of 1917, Aberdeen & Kincardine Area had 15 Branches. In the minutes which marked the formation of Stonehaven Branch in 1917, William Hunter, Redcloak was listed on the committee. He became convener of the Union's Labour committee and was still to the fore at the Union's 50th anniversary.

Also in November 1917, a Branch was formed at Insch where one of its first tasks was campaigning for the re-opening of the Insch Mart, which had been closed earlier that year. That very same month, the Union spread its influence north by the formation of a Branch in Turriff. One of its first committee members was J A D Pirie, Castle of Auchry who went on to become a president of the Union. Another from that inaugural meeting in Turriff was W Norrie, Cairnhill whose son A G Norrie went on to be convener of the Poultry committee and whose grandson was president of the Turriff Branch in 1964. Inverurie Branch was formed in 1918 and one of its founder members was Dr James Durno of Uppermill; he was one of Scotland's best known and respected agriculturalists. Garioch disappeared into Inverurie & Oldmeldrum in 1920 and Cluny & District was broken up in 1924. Then Crathie, Ballater and Braemar went out of the records in 1934. Earlier Glenkindie came under the wing of Alford in 1917. Then during reorganisation in 1967 the Branches of Birse and Cromar formed Mar Branch.

In Banffshire the Branches at Banff and Keith were both formed at the end of 1917. At this time Banffshire came under the aegis of Aberdeen and Kincardine and only became an Area on its own in 1943. Marnoch Branch came into existence in 1947. At reorganisation in 1967, Rathven and Deskford were joined to form West Banffshire.

The first available minutes from Moray date from 1927 but this Area was formed with one Branch at Elgin in 1918, at the same time as the adjoining Area in Nairn was formed. In 1927 Moray Area spread its wings to envelope Branches at Forres and Strathspey and in 1933 Knockando & Elchies Branch was set up bringing Moray's Branches up to four.

By 1990 Aberdeen & Kincardine had 1,680 members in 15 Branches; Banff had 468 members in six Branches, while Moray and Nairn had 312 members in four Branches. Following regionalisation in 2000, the former North East Areas were left with 11 Branches. These were Huntly & Insch, Stonehaven & Banchory, Laurencekirk, Turriff, Ellon, New Deer, Buchan, Aberdeen, Banffshire, Inverurie & Oldmeldrum and Dee/Don.

John Cameron addresses an Inverness area meeting in July 1986, along with from left to right, Murray White, area secretary; Lindsay Girvan, area president; and Jock Hunter

Secretary of Banff Area for 32 years, Alan Meldrum remembers a number of the North East personalities: "Alan Taylor, Ordens was a strong supporter of the Union. He helped build up a strong Area executive committee and also served at headquarters as Pigs committee convener. In addition he was chairman of North Eastern Farmers co-operative. Alan Taylor Jnr, Boghead followed his father in continuing and progressing the Union image, particularly helping with recruitment to nearly 100% membership. He served on the General Purposes committee and also on the Scottish board of the Mutual.

"John Robertson, Renaten farmed single handed and yet also devoted days of his time in furthering the case for hill and upland farmers. It meant he had to be up early to set off for Council and committee meetings, attending to stock before leaving and then again on returning home late in the evening.

"Dick Hillson was secretary for Moray and Nairn before becoming Organiser of the Union and editor of the Farming Leader. I recall a true story about Dick who had put on a bit of weight. He had a fondness for chocolate and sweets. The result was that his weight rose at one stage to more than 18 stones and, on calling on a farm in Speyside, the farmer's wife remarked on this weight increase. Dick acknowledged that he had and was up to 18 stones. The good lady stopped him in his tracks saying: 'My goodness, you're nearly a ton!' As a noted bon viveur Dick used to entertain at the Union's after dinner session at the annual meeting by singing - A lum hat wantin' a croon."

Former president Sylvie Campbell recalls other North East stories: "Once after a Branch meeting where the BBC was strongly criticised for putting a wrong slant on the costs of producing the nation's food to the general public, the secretary was asked to make farmers' annoyance very clear to that body.

"Bits of humour did occasionally surface at Branch meetings. At one meeting we were being talked to by a soil scientist from the Macaulay. He showed a slide of a deep hole, which had been dug on a local farm showing beautiful soil right to the bottom. A voice was heard at the back of the hall, 'No wonder Jock can get fu' every night when he has soil like that.'

"As a young man attending my first annual dinner when the whisky was flowing rather freely, Bertie Maitland advised me, 'Just put a suppie milk in it.'

"I suspect that Branch and Area were no more strongly led in the early days by men like Capt Manson of Kilblean and J Stephen of Conglass than they are now by present office holders but it seems to me that member backing was stronger. There were more Branch meetings, more Area meetings but then there were more farmers. One hundred acres used to provide a living of sorts for a family and one or two workers but not nowadays."

Red Tape and Supermarket Power

In many ways the most recent events in Union history are the most difficult to record. A passage of time is often needed to identify which issues have been significant and which were ephemeral. The following are the main themes the Union has had to deal with in modern times but other problems especially specific animal diseases may make a bigger impact even if they currently seem marginal.

Purchasing Power

One of the primary motives in establishing the Union in the early years of last century was to improve the producers' share of the money paid by the customer for milk. This issue rose and fell in the years that followed but the 1990s and the first decade of the twenty first century saw Scottish farmers organise and take part in a spate of demonstrations and protests over their poor share of the cash in the food chain.

In the majority of cases taken up by the Union, the main targets for the protests were the major retailers. In the second half of the last century, the country had experienced a major retailing revolution with the demise of many small, local shops and the rise of a few, very large and therefore very powerful supermarkets. These massive retailers then used their purchasing muscle and cut the price paid for milk, potatoes, beef, lamb and every other foodstuff to their suppliers. To highlight the imbalance in the food chain, the Union in 2005, highlighted the gap between the prices paid to the primary producer and the prices the produce was selling for on the shelves of the major multiples. Milk which had been bought at 18 pence per litre was being sold at 54 pence, potatoes which had cost £95 per tonne were on sale on the supermarket shelves at the equivalent of £550 per tonne, growers of oats were being paid £70 per tonne while the same oats by now packaged as porridge oats were on sale for £570 per tonne, beef at £1.90 per kilo to the farmer was transformed into beef at £4.25 per kilo by the time it was offered for sale on the retailers' shelves and strawberries for which the producer was paid £1.30 per pound became worth £3.00 per pound through their short trip to the supermarket.

Recognising something serious had to be done about the imbalance, the Union pressed for a fair trade deal and pushed for the Code of Practice which had been brought into being in 2002 to be strengthened. This Code, the Union argued, was

Milk committee chairman, Robert Shearlaw (front centre) with other members of the committee outside Westminster before addressing MPs on the action needed to save Scotland's dairy industry and protect rural communities

"too woolly" and was only relevant for those dealing directly with the major retailers; thus farmers were outwith its remit. Delegations of Union members lobbied Westminster and Holyrood highlighting some of the practices of the supermarkets.

In that summer of farming discontent in 2005, the Union took the discrepancy in price right to the doors of the supermarkets with members coming face to face with consumers in putting the case for a fairer food chain.

These demonstrations, although short lived and local in nature, provided a major breakthrough in re-establishing a link between producer and purchaser; a link that had slipped off the horizon in the 1960s and 1970s with buyers often having no awareness of the origins of the food they bought.

Then in 2007, the Union brought forward a campaign aimed directly at informing consumers of where food came from. It had a title, 'What's on your

NFUS members speak to shoppers outside an Edinburgh supermarket

The NFU Scotland board of directors enjoy some quality Scottish fare at the launch of the 'What's on your plate' campaign

plate' and it was aimed specifically at making the public think about the provenance of the food they were buying. One example of the many local actions on this campaign saw Ayrshire members attend the local Farmers Market in July 2007 to spread the word on the campaign. The report from this direct action was positive with the members stating there was no difficulty in getting the locals of Ayr to give their full backing on the day.

The appeal to consumers to check where their food came from has been one of the latest successes for the Union as it has helped nullify part of the power of the supermarkets. Another more recent positive move came in 2013 when the UK Government appointed Christine Tacon as Groceries Code Adjudicator. Her remit will involve ensuring fair play between the nine major retailers and their direct suppliers.

The purchasing priorities of the major retailers had been seen earlier in 1997 when members in the south west of Scotland observed an increasing number of Irish lorries filled with meat coming in via Stranraer. While the Leader reported that the "Union has just staged a peaceful demonstration to show the public and Government that we need justice for our industry," the reality was that the issue saw the resignation of both the Union president and the chief executive as members felt the Union was not taking a sufficiently high profile on the issue. The imported meat was subsidised by the Irish Government and Scottish farmers felt they were fighting with their hands tied behind their backs. The Union took the issue to London but the anger did not finish there as a mass rally along Princes Street in Edinburgh in January 1998 demonstrated.

Four years later, Union members were back in Edinburgh demonstrating but this time, they had a different target. Diageo, the international drinks company were importing barley to make Scottish whisky and were not prepared to give growers

The Union took the fight against imported meat to London and then organised a mass rally along Princes Street in Edinburgh in January 1998

in this country, forward contracts for malting barley. Union president Jim Walker claimed the gamble farmers had to take on growing malting barley would backfire on the whisky industry one year when the growers decided not to go for the malting market.

Red Tape Battle

The battle for a fair share of the food chain was one of the main themes of the first decade of the twenty first century but another major issue for the farming industry in that same period was highlighted in the June 2006 edition of the Farming Leader. The article was headlined 'Drowning in Regulation' and it listed the latest Directives and Regulations emerging from Brussels along with the actions the Union was taking to mitigate the bureaucracy faced by the farming industry. Red tape was not an issue that arrived suddenly on the doorstep of the Union with complaints over excessive bureaucracy being a regular agenda item on Union meetings right from the early days of last century.

The more recent appearance of compulsory form filling as a contentious issue for the

Message in a bottle. Empty whisky bottles pile up outside Diageo, all containing a message demanding fixed price contracts

industry was the sheer volume of regulations emerging from Brussels. The European Union had developed from its early days where production was paramount to one where it increasingly became involved in regulation and control; thus introducing new legislation.

The tendency for this country to 'gold plate' legislation has increased the workload of the Union in recent years. In 2008, Union chief executive, Andy Robertson claimed there were 2,281 pieces of European legislation relating to agriculture with another 568 relating to the environment. He proposed a number of points to be considered before any further regulation was introduced. These were: Was there a problem and was there an alternative to regulation? Robertson also proposed that new regulations should be targeted in priority areas and a cost/benefit analysis to be carried out before implementation. Financially he said that any charges linked to the new regulation had to be justified by open transparent calculation and finally all regulation had to be subjected to review after implementation.

One of the first and also one of the longest battles on legislation the Union has had with Europe has been over Animal Transport; particularly on how many hours livestock should be in transit. Legislation limiting travel time was first mooted in 1995 with the Union stating this might be desirable and necessary on mainland Europe where long distances can be covered on a modern motorway network but restrictions on essential crossing of animals from an island or a remote part of Scotland to a market would disadvantage famers and crofters.

In a more recent example of European politicians not understanding the Scottish system of farming, the Union along with the National Sheep Association has battled legislation on the electronic identification of sheep. As a legacy from the 2001 foot and mouth outbreak where the scale of sheep movements was realised too late in the day, this was introduced to help monitor the movement of sheep. Arguments that sheep born and kept on units would not need to be tagged until their move off farm failed to convince the Brussels legislators.

Another long standing and continuing issue for the Union has been the introduction of Nitrate Vulnerable Zones (NVZ) which were brought in by the EU as part of plans to reduce nitrites in ground water. For Scotland the designated areas are mostly in the east or arable side of the country where they restrict the spreading of farm yard manure and limit applications of nitrogenous fertiliser. But being in a NVZ also affects livestock farmers. In 2007, farmers in the NVZ Nithsdale catchment area worked out that the cost of storing slurry for 6 months rather than the previously permitted three month had put another £1,000 on their expenditure.

In another dispute over water, the Union took up the issue of farmers who used irrigation for their potato, soft fruit and vegetable crops being charged for the water abstracted. Under the EU Water Framework Directive, those farmers required a licence and also required to pay for the amount of water they took. The Union argued that this was a financial penalty that went against local food production.

They lost this battle but gained some ground in removing some of the initial proposals from the Scottish Environment Protection Agency. One example was a proposal to put a £3,000 price tag on each abstraction point but this was removed following Union intervention.

Scottish Legislation

Adding to the pile of legislation and regulation facing the Union in the past 14 years has been the newcomer on the governance block, the Scottish Parliament. Scottish Parliamentarians wasted no time in flexing their legislative muscle in trying to catch up on three hundred years where Westminster had been responsible for legislation governing Scotland.

High on the agenda for Scotland was a wide ranging look at land ownership, farm tenancies and access to the countryside. One outcome from this review saw, in 2000, a landmark agreement being struck between the Union and the Scottish Landowners Federation on a new basis of letting land in Scotland. Two distinct types of Limited Duration Tenancies were introduced. They were designed to address separate markets; the first lasting for up to five years for cropping arrangements; the second for at least fifteen years was for longer whole farm lets.

At the beginning of the last century, the vast majority of land in Scotland was farmed using the tenancy system but many large estates were broken up post the First World War in order to pay Death Duties. In an earlier effort to put some stability into the tenanted sector, the Union was heavily involved in the 1948 Agricultural Holdings Act which moved tenancies from their previous seven or fourteen year lease into lifetime agreements.

Despite that ground breaking agreement and the later one in 2000, the number of tenancies and the percentage of Scotland under tenancy agreements has continued to fall to a point where now only one third of farmed land in Scotland is tenanted. The Union hoped the 2000 agreement would initially stabilise and subsequently increase the tenanted sector but one of the big fears for landlords letting out land has been the threat of tenants being given an absolute right to buy the farms they are leasing; a threat that was first raised in the Scottish Parliament in 1999 and which continues to the present day.

Tenancy issues were not the only concern arising from new legislation produced by the Scottish Parliament as the Land Reform Bill which came out in 2001 gave more access to the countryside. Union president, Jim Walker commented, "Farms are homes as well as being places where people earn their livelihood. The Bill has given some protection to these roles by keeping farm steadings private and recognising that hay and silage are crops which need to be protected."

The demise of the Marketing Boards

By the mid 1990s, the threat to the future of two of the remaining marketing boards – milk and potatoes – turned into reality with the UK Government deciding the statutory powers held by the boards did not chime in with their policies of open and competitive marketing.

The Union fought tenaciously to keep them both. The battle for the retention of the three Scottish Milk Boards began in 1978 when Europe decided they did not fit in with European policies. The Union organised a poll of the 3,196 milk producers in Scotland at that time.

Of that number, 3155 voted with 3140 supporting the retention of the boards. Commenting on the result, Henry Christie Union Milk convener who had spent a considerable amount of time and effort going round milk producers in Scotland holding 'byre meetings,' said "We asked for overwhelming support and that is what we got." The positive result also produced support from the Scottish Secretary of State, Bruce Millan, who said the poll result "confirmed in a resounding way, the confidence of producers in their Boards."

However that poll merely provided a stay of execution with the next attack on the boards coming in the early 1990s from the UK Government who wanted to implement the de-regulation of the milk market. John Selwyn Gummer, as UK Agricultural Minister, and Lord Strathclyde, as Scottish Secretary with responsibility for Agriculture north of the border, wielded the sword leaving the Union to, once again, defend the milk boards. At the height of the dispute over the future, president John Ross recalled an incident one Saturday afternoon in Stranraer. "The Thatcher Government was determined to de-regulate the milk industry. A group of dairy farmers who were strongly opposed to this change discovered that Ian Lang, MP for Galloway and a member of Mrs Thatcher's cabinet, was due to visit Stranraer on constituency business. Despite him explaining to the group that he had urgent business to attend to, the farmers detained him in a car park for some considerable time until he was fully briefed on their concerns."

However despite the Union's best efforts, the milk industry was subsequently deregulated in 1994; thus closing down 61 years where milk marketing had been backed in Great Britain with statutory powers.

Prior to the demise of the milk boards, a poll of milk producers had backed the creation of a Great Britain wide Milk Development Council which would continue some of the promotional activities of the milk boards.

As soon as it was obvious that the writing was on the wall for the statutory Boards, the Union also actively campaigned for the establishment of a member owned co-operative to match the competitive challenges which would come post-regulation. They were successful with the establishment of Scottish Milk (now First Milk) but ever since the demise of the boards and control of the processing factories they owned, the financial imbalance in the milk supply chain has been the trigger for a number of major demonstrations by Union members angry at their share of the final milk price.

Following a number of protests at the distribution depots in 2001, the following year saw one of the most successful demonstrations in the Union's history, when almost a third of all Scotland's dairy farmers turned out at Asda's distribution centre in Grangemouth. As a result of this action, the Union secured a commitment from the major retailers for a two pence per litre increase for dairy farmers selling milk for the production of cheese and other dairy products.

In 2002, NFUS president, Jiim Walker, addressed the NFU demonstration at the ASDA distribution centre in Grangemouth at the height of the battle to increase the milk price. The demo which comprised nearly a third of the industry proved a turning point. NFUS secured at least 2p per litre rise

A decade later in July 2012, dairy farmers from all over the UK met at Lanark Market as part of a campaign to stop processors reducing the price they paid for milk. There they heard Union president Nigel Miller support plans for Producer Organisations to be established to help counteract the power held in the hands of a few processors and retailers.

This campaign also saw a major rally being held in Westminster with UK Farm Minister, Jim Paice very much in the firing line. With producers also getting consumers on their side through actions on the doorsteps of several major retailers, the processors backed off their price cut and the UK Government pushed ahead with a Voluntary Code of Practice. The campaign had been successful and the Union had played an important role in it. In Scotland, Scottish Government Rural Affairs Minister, Richard Lochhead promised a review into the future of the milk industry. The publication of this document in 2013 painted a bright future for the 1,000 milk producers left in Scotland.

The Potato Marketing Board suffered a similar fate to the milk boards with political opposition emerging in the 1990s to its attempts to control the market through limiting the acreage planted. After the Government announced in 1994 that the PMB statutory powers would be removed in three years time, the Union stressed the need for rapid progress towards Development Council status for its successor body. When the British Potato Council took over from the PMB on 1st July 1997, it was agreed the new body would also take over the work of the Scottish Seed Potato Development Council.

Changes in the Common Agricultural Policy

By 1999, the Common Agricultural Policy budget was split into two pillars. The first of those was the recognisable direct support to producers which, at that time, was based on numbers of cattle and sheep kept and acres of cereals grown. The second pillar for financial support from Europe was rural development. As this included Less Favoured Area support, it was vitally important for Scottish agriculture.

But it was the next reform of the CAP under the guidance of EU Commissioner Franz Fischler that brought in a more fundamental change. At their meeting on 26th June 2003, EU Farm Ministers adopted a policy of 'decoupled' subsidies so that the CAP would not infringe World Trade Organisation policies against production based subsidies. The new Single Farm Payments were subject to "cross-compliance" conditions relating to environmental, food safety and animal welfare standards. In their own thorough consultation on the proposals which were due to come into force in January 2005, the Union fully supported full decoupling in the arable, beef, dairy and sheep sectors. The Farming Leader did register a caveat that "the suckler herds would have to be watched particularly in the more fragile areas." Taking this caution, the Scottish Rural Affairs Minister Ross Finnie introduced a Beef Calf subsidy in an attempt to retain beef production in Scotland. Union members said the SFP must be calculated on the basis of individual, historic entitlements during the period 2000-2002. Few forecast the danger of paying subsidies for non production; an issue which in 2009 saw Richard Lochhead, Cabinet Secretary for Rural Affairs in the Scottish Government visit the Isle of Mull at the behest of Argyll regional board chairman Bert Leitch.

There the politician saw the negative impact that falling numbers of livestock and land abandonment could have on a rural area with Mull having lost a significant proportion of its breeding stock leaving large areas of the island no longer being grazed.

Land abandonment and inactive farmers have become a core issue for the Union in the most recent CAP reform due to be implemented in January 2015.

Other features of the latest CAP reform will see reduced levels of direct support being paid on an acreage basis with this money being paid only to those farmers compliant with complex European wide environmental regulations; this linking of farming and the environment being a primary aim of Commissioner, Dacian Ciolos.

Miscellaneous

Secretarial Help In 2001: NFU Assist was launched with the aim of providing a secretarial service that would "Take all the worry and hassle away from the farmer, freeing up time to concentrate on the main tasks on the farm." The scheme, the first commercial venture by the Union, died within two years.

Fox Hunting Ban: In 2001 the Union gave evidence to the Rural Affairs Committee of the Scottish Parliament on a Bill which sought to implement a ban on hunting with dogs. The Union made the point that the average sheep farmer would have to survive on £700 per year and fox predation in some remote areas could reduce the lambing production to just 50%. Any attempts to restrict the rights of farmers and crofters to protect their assets had to be resisted according to the Union. The Bill was promoted by Mike Watson and was passed into law on 13th February 2002.

Pesticide Tax Victory: The combined lobbying power of the UK farming Unions and the Crop Protection Association helped avert the imposition of a Pesticide Tax in the early years of the twenty first century. The industry position was that a series of proposals and commitments for voluntary measures such as sprayer operator courses would minimise the impact of pesticides on the environment. However, there was less success for the Union in 2009 when the EU banned a large number of pesticides; a pressure that continues to the present day.

Sea Eagles swoop: The re-introduction of sea eagles in Scotland has been hailed a success by environmentalists but by 2008 they were already causing grief to hill sheep farmers and crofters. Highland Region Secretary Ian Wilson stated after attending a meeting about sea eagles in Poolewe, "The damage they are causing to sheep flocks in that area is a real concern for the crofters and farmers who can do nothing to protect their lambs. One concern to me seems to be the unwillingness of RSPB and Scottish Natural Heritage to believe the level of damage done by the sea eagles."

The union, as an organisation

Union moves
For the past twenty one years, NFU Scotland headquarters has been based at the Rural Centre, Ingliston pic where it shares the premises with a number of other rural organisations including Quality Meat Scotland, SAOS – the umbrella organisation for co-operatives and RSABI – Scotland's rural charity. The Rural Centre site is also adjacent to the Ingliston showground where the annual Highland Show is held.

The Rural Centre, adjacent to Ingliston Showground, near Edinburgh, is the current home of NFU Scotland

Working in proximity to those rural organisations provided one incentive for the move but another one was the increased difficulty for the Union of operating out of offices in Grosvenor Crescent in the New Town area of Edinburgh. The Union had bought the Grosvenor Crescent offices in 1946 paying the sum of £3,000 for the listed building. Initially delegates attending committees often came by train to the nearby Haymarket Station but increasing use of cars brought its own problems with parking; especially after the installation of meters outside the offices. Delegates either arrived with a pocket of cash to feed the meters throughout the day or left with a parking ticket.

The Grosvenor Crescent office was the first property owned by the Union as it had previously rented accommodation in Ainslie Place, Edinburgh following a transfer from Glasgow in 1936; the minutes recording this move across country as being necessary "on account of the increased volume of work at headquarters."

The original base of the Union was in St Vincent St, Glasgow where A Hunter, the Union's first General Secretary worked as a lawyer.

Social side of the Union
The vast majority of Union effort is work related but the annual meeting along with the dinner has now woven itself into part of the social fabric for many

Union members. While, as a properly constituted organisation, the formal annual meeting has been there from the start, the social side of the annual dinner has only emerged since the 1950s. In the early days, the dinner preceding the annual meeting was an informal affair. It was more a meeting for those who had travelled from a distance the previous day but it has now developed into a major social function with upwards of 500 members, spouses and corporate guests attending. Again in the early days, the annual meeting was either held in Glasgow or Edinburgh but by the 1960s, it was being held in a different area every year.

As befits an industry producing food, the dinner menu has become a major feature of the occasion. In 1964 at the fiftieth annual meeting held in the Turnberry hotel, Scottish food featured for the first time, albeit prepared by the hotel's French chef, Monsieur Cottet. The menu was Ayrshire hors d'oeuvres for a starter, Oban scampi Jubilee for the fish course, roast Shetland lamb with stovies, roast potatoes and garden peas as a main course, followed by soufflé Ailsa Craig and a Scottish cheese board.

Locally sourced food has become a mainstay of the NFUS annual dinner nowadays. The event has developed into a major social function in the agricultural calendar

With the Union touring around the Areas for its annual meeting, it soon followed that the food for the dinner was locally sourced. The trencherman behind this initiative was John Leese, a long time supporter of Scottish food. In 1974, Orkney provided the produce for the menu; a first for the Areas and Regions and a practice that holds to the present day even, although the event has been held in St Andrews for the past several years.

Publicity and the Union

The first Union president, William Donald, Fardalehill, Kilmarnock recognised that an essential component of a successful lobbying organisation was positive press coverage. On 28th November 1919, he urged the Central Executive committee to appoint a "Press Propaganda Secretary" to combat the derogatory statements about farmers appearing in the newspapers. After discussion this proposal was unanimously agreed.

By 1933, Branches were being urged to send news material to the North British Agriculturist and Farming News magazine where the Union had made arrangements for a whole page to be devoted in this weekly paper to NFU matters. In reporting this news, East Lothian Branch urged its members to assist by writing articles on subjects pertaining to agriculture. These articles had to be forwarded either direct to newspaper or to the General Secretary.

World War Two brought severe rationing of paper and the Union had to cut back on its publicity work. However, by 1948, a fulltime publicity officer was appointed by the Union to work under the direction of the Publicity Committee. The minutes also report that a "propaganda" leaflet describing the aims, purpose and function of the Union had been prepared. A couple of year later, it was reported that there had been an easing of newsprint supplies and as a result, agricultural and Union activities were able to claim more space in press columns.

At the same time, the Union pushed ahead with its own members' magazine, the Farming Leader. This was first published in January 1961 and has appeared monthly ever since then even although concern was expressed over the £800 per annum it originally cost to post it to all members.

Also in the 1950s, the Union noted its pleasure over the increasing coverage of agricultural topics by the BBC. This was the first reference to radio and television coverage which soon became an integral part of media conferences, especially when there were food scares. It was no surprise in February 1996 when Tom Brady proposed the setting up of a Public Affairs Division at headquarters. The key objective of this new division under Polly McPherson, was to "improve the public image of farming and the rural community, to improve and co-ordinate the Union's message when under attack on food, welfare and the environment and generally to improve internal and external communication.

Communications today would leave the founders of the Union gasping with extensive use of texts, emails and blog sites to communicate issues news and events far faster than ever before.

Internal organisation of the Union

Although the basic working framework of the Union with its Branches and Areas has remained in place for the past century, there have been several other major changes in its structure and organisation. Prompted by a letter from Ayr Area sent in October 1975, the Union set up a special committee under the chairmanship of Sandy Inverarity to look at the whole operating structure of the Union. Among the many recommendations that this committee proposed and which were accepted by the Union was one to move away from presidents only serving one year in office. A longer term of office was needed to take on the increasingly complex issues coming from the Common Market, it was argued although a survey of the twenty previous presidents found that only three of them would have been prepared to stand for more than one year in the top office. The English Union had adopted the practice of longer term presidents just after the Second World War thus explaining why it reached its one hundred years anniversary in 2009 with only 33 presidents compared with the Scottish Union having 60 in that post by its centenary year.

The special committee also recommended further reducing the number of Branches as the number of farmers were falling as farms increased in size. Branch numbers had peaked at 173 in 1963 but some rationalisation had taken place. The committee suggested a network of 130 could cover the whole of Scotland effectively without members having to travel too far.

A third recommendation by the committee saw the rejection of a proposal to have a tourism committee for the Union based on the large numbers of farms now diversifying into having bed and breakfast operations or farm shops. The committee felt that there were other existing organisations to help those working in tourism. However they did recommend adding a Fish Farming committee, a Glasshouse and Nursery committee and a Soft Fruit and Field Vegetable committee.

The committees of the Union have come and gone as the workload has presented itself. As has been noted, the Milk committee is the sole survivor of the 13 committees set up in the beginning. Others such as the Livestock committee started life as the Beef and Mutton committee.

Some including Game Laws, Local Taxation, Parliamentary Representation, Heather Burning and Deer Forests have either slipped off the agenda altogether or seen their work subsumed into today's committees which include Environment and Land Use, Less Favoured Areas and Legal and Technical.

The rest of the 2013 committees cover Combinable Crops, Pigs, Potatoes, Poultry, Soft Fruit and Field Vegetables, Specialist Crops and Tenants.

Possibly the biggest internal change to the Union came after a Council meeting in 1947when it was unanimously decided "it would be in the best interests of the Union to arrange for the appointment of whole time Group Branch Secretaries." There was no timescale for this proposal with the minute recording "as opportunities arose" but by 1952 the Organisation committee of the Union gave favourable consideration to a request from Banff Area executive that a whole-time group secretary should be appointed for the County. Arrangements were made for the post to be filled and Mr J Cruickshank, Keith, was appointed. This, the minute records brought the number of whole-time Group Secretaries to four; appointments having been made previously in Angus, the Borders and the Moray and Nairn Areas.

By 1957, the Union expressed the view that the whole country should be covered by full time Area secretaries within the next ten years, and although that decade target was missed, by 1969 the objective was achieved and that position has been retained to the present day.

NFUS
A dozen achievements

■ Establishment of marketing boards in 1933 following years where imports had almost crushed the home industry. Less than a decade later, the importance of home production proved critical in winning the Second World War.

■ Campaigns against a range of animal diseases. Bovine Tuberculosis was a major problem in the first half of last century but Scotland had it eradicated by the mid 1960s. Aujesky's disease in pigs was successfully eradicated in the 1980s. Sheep scab was cleared out in the 1960s only to come back in again in more recent times. The 2001 Foot and Mouth epidemic was the worst ever animal health problem to hit this country and the Union played a major role in controlling and defeating it. BVD, Johne's and Schmallenberg are all currently being tackled.

■ While it has no control over the weather, the Union has on numerous occasions in the past one hundred years either co-ordinated relief operations in disaster hit areas or it has helped gain compensation for climatic extremes.

■ The Union has ensured that Scotland's hill and island areas were recognised when the UK entered the EU and has continued to fight for those farming in the less favoured areas of Scotland

■ For twenty years from the 1970s to the 1990s fought the injustice of the "green pound" disparity which saw UK producers get 40% less for their pigmeat, cereals, milk and beef compared with European producers.

■ Fought the French over their illegal ban on Scottish sheep meat. This was a long but ultimately successful campaign.

■ Highlighted the damage caused to the home industry by imported beef from Ireland and from South American countries. This was initially through demonstrations at Stranraer and Merklands and latterly across the negotiating table.

■ Highlighted the issue when the distillers imported foreign barley to make Scottish whisky. The most recent example being in 2002 but the first was in 1925 when Nairn County wrote to headquarters asking for a tax to be imposed on all foreign barley delivered at distilleries.

■ Along with other Unions in the UK, NFUS set up an office in Brussels forty years ago to ensure Scottish matters were considered in all European Union discussions.

■ Took a primary role in negotiations with Government on farm policy. This was first established in 1945 with the annual price review and continues to this day with meetings with both UK and Scottish Governments

■ Took a primary role in promoting Scottish food in the late 1970s; a stance which has now blossomed into the positive promotion of produce grown in this country through a number of organisations, including Quality Meat Scotland and Scotland Food and Drink

■ Successfully unified a geographically disparate industry with numerous competing enterprises and thus confounded the doom merchants who believed farmers would never cooperate.

A century of leaders

Below is a list of Presidents, Vice Presidents and Chief Executives who have dedicated their time and expertise toward making the National Farmers Union of Scotland what it is today.

President	From	Date
W Donald	Ayrshire	1914 – 1919
J Gardner	Renfrewshire	1919 – 1921
J Lennox	Perth	1921 – 1924
W J Dudgeon	Sutherland	1924 – 1925
A Batchelor	Forfarshire	1925 – 1927
J Speir	Lanarkshire	1927 – 1928
Sir J P Ross-Taylor	Mid & East Berwick	1928 – 1930
W Cassels Jack	Lanarkshire	1930 – 1931
M Mackie	Aberdeen, Banff & Kincardine	1931 – 1932
W Bruce	East Lothian	1932 – 1933
G G Middleton	Black Isle	1933 – 1934
W Hutcheson	Roxburgh	1934 – 1935
J A D Pirie	Aberdeen, Banff & Kincardine	1935 – 1936
G Buchanan	Renfrew	1936 – 1937
J Picken	Stewartry	1937 – 1938
A Manson	Aberdeen, Banff & Kincardine	1938 – 1939
W J Wright	East Lothian	1939 – 1940
W Graham	Kelso	1940 – 1943
W D Jackson	Peebles	1943 – 1944
J Rennie	East Lothian	1944 – 1945
W Young	Ayr	1945 – 1946
H C Falconer OBE	Mid & East Berwick	1946 – 1947
I M Campbell	Sutherland	1947 – 1948
D Lowe	East Lothian	1948 – 1949
A H B Grant	Perth	1949 – 1950
A R Semple	Dumfries	1950 – 1951
J C Wallace Mann CBE	Mid & West Lothian	1951 – 1952

President	From	Date
G Hedley	Peebles	1952 – 1953
J Leiper	Rhins of Wigtown	1953 – 1954
J Marshall CBE	Perth	1954 – 1956
H D Brown	Ayr	1956 – 1957
J E Rennie	East Lothian	1957 – 1958
J Johnston	Stirling	1958 – 1959
A R J Milne	Aberdeen & Kincardine	1959 – 1960
J G Jenkins	Dumfries	1960 – 1961
W B Swan CBE	Mid & East Berwick	1961 – 1962
D Goodfellow CBE	Angus	1962 – 1963
I A Campbell	North Argyll	1963 – 1964
M Joughin CBE	Moray	1964 – 1966
W W W Peat CBE	Stirling	1966 – 1967
C Young	Angus	1967 – 1968
J L Blackley	Dumfries	1969 – 1969
J A McIntyre	Rhins & Wigtown	1969 – 1970
J A Inverarity	Perth	1970 – 1971
D M Milne CBE	Angus	1971 – 1972
A A Arbuckle OBE	East Fife	1972 – 1973
J Stobo OBE	Mid & East Berwick	1973 – 1974
S Campbell OBE	Aberdeen & Kincardine	1974 – 1975
F R Evans	Wigtown	1975 – 1976
W P Watt	Banff	1976 – 1977
M R Burnett	Sutherland	1977 – 1979
J B Cameron CBE	Fife & Kinross	1979 – 1984
I D Grant CBE	Perth	1984 – 1990
J A Ross CBE	Wigtown	1990 – 1996
G A Mole	Borders	1996 – 1997
G Lyon	Bute	1997 – 1999
J L Walker CBE	Dumfries & Galloway	1999 – 2003
J Kinnaird	Lothian & Borders	2003 – 2007
J C McLaren	East Central	2007 – 2011
N A Miller	Lothian and Borders	2011 -

Vice presidents

The vice presidents listed below gave leadership to the Union, valuable assistance to the president, but for one reason or another so far, have not taken the final step toward the presidency.

President	From	Date
C H Beveridge	Elphinstone Tower, Tranent	1929 – 1931
T P Burnett	Newton, Methlick	1929 – 1931
T C Lindsay	Aitkenbrae, Monkton	1929 – 1930
D McLaren	Bracklinn, Callander	1929 – 1930
A Munro	Leanach, Culloden Moor	1929 – 1930
D E J More	Dalmacoulter, Airdrie	1930 – 1932
G G Mercer	Southfield, Dalkeith	1932 – 1933
R Garvie	Hillocks of Gourdie, Blairgowrie	1936 – 1937
J Vallance	Auchness, Stranraer	1938 – 1939
J Russell	Walston Mansion, Dunsyre	1939 – 1940
W Hunter	Redcloak, Stonehaven	1940 – 1942
G Dunlop	Baltersan, Newton Stewart	1964 – 1965
H B Christie	Myrton, Newton Stewart	1977 – 1982
J B Forrest	Whitemire, Duns	1979 – 1980
J P S Hunter	West Highland Estates Office, Fort William	1980 – 1981
G A B Anderson	Kair, Fourdoun	1981 – 1986
J W Hay	Carlungie, Carnoustie	1984 – 1986
J L Goodfellow	Cairnton House, Arbroath	1986 – 1989
M Mackie	Westertown Farm, Rothienorman	1989 – 1994
J Wyllie	Ruchlaw Mains, Dunbar	1994 – 1996
S W Whiteford	Rarichie, Nigg	1996 – 1998
P Chapman	Peedie House, Strichen	1998 – 2000
P Stewart	Urquhart Farm, Dunfermline	1999 – 2003
R Howat	Drycleuchlea, Hawick	2003 – 2007
D Mitchell	Pairney Farm, Auchterarder	2003 – 2006
S Wood	Garson Farms, Orkney	2007 – 2009
A Bowie	The Haining, St Andrews	2010 – present
J Picken	Priorletham, St Andrews	2011 – 2013
R Livesey	Firth, Melrose	2013 – present

Secretaries/Chief Executives

Name	Date
A W Hunter	1929 - 1930
Miss A MacLaren	1931 - 1943
W Graham, C.B.E.	1943 - 1955
H G Munro, LL.B., W.S.	1955 - 1978
D S Johnston, B.A.	1978 - 1996
T Brady, M.A.	1996 - 1997
E Rainy Brown	1998 - 2003
A Robertson	2003 - 2008
J Withers	2008 - 2011
S Walker	2011 - present

Sandy Inverarity, cutting the cake at the centenary reception in October 2013

Acknowledgments:

Sincere thanks to the following individuals and organisations for their help in sourcing material and images that have been incorporated into this book.

Peter Small pictures on pages 9 and 34.

Orkney Library and Archives reproduction on page 15.

Midlothian Council Local Studies picture on page 17

The Scotsman Publications Limited picture on page 23.

East Lothian Library Service picture on page 36.

National Museum Scotland pictures on pages 37 and 90

The Scottish Farmer pictures on page 43

Hilary Barker pictures on page 71

The Courier, Dundee, picture page 117

Front cover image of The Carse of Stirling. Photographer, Wayne Hutchinson